DEADLY SUNSHINE

THE HISTORY AND FATAL LEGACY OF RADIUM

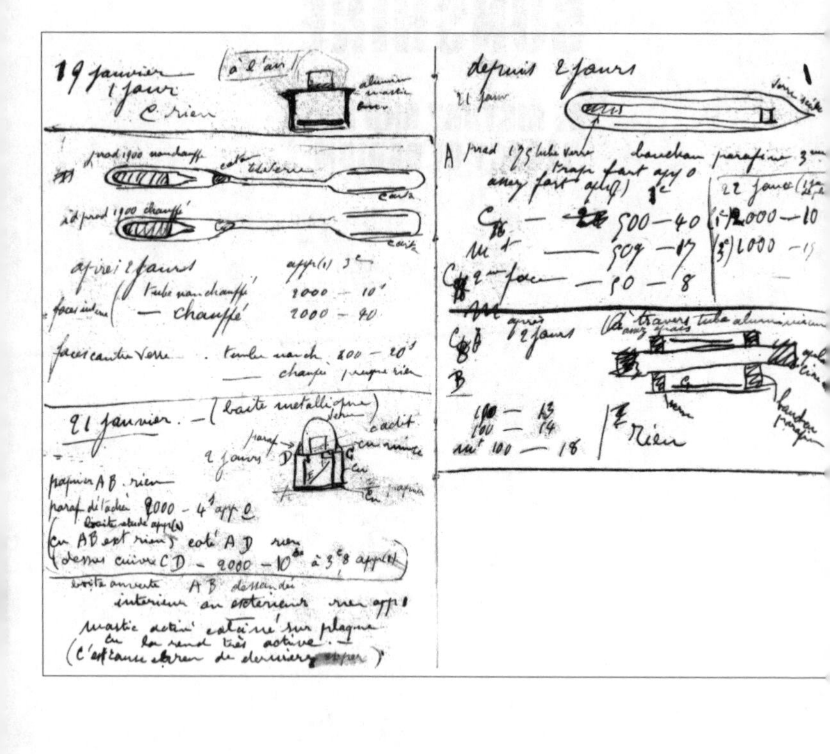
19 janvier
1 jour
C rien
(à l'air)

2d prod 1900 chauffé

après 2 jours
facesinternes (tube non chauffé 2000 – 10
— chauffé 2000 – 40
faces contre verre . tube non ch. 200 – 20
— chauffé presque rien

21 janvier. — (boîte métallique)
2 jours
papier AB . rien
paraf. détachée 2000 – 4 app 0
(en AB est rien) côté AD rien
(dessus cuivre CD – 2000 – 10 à 3 app
intérieur ou extérieur rien app
mastic activé catalysé sur plaque
la rend très active. —

depuis 2 jours
21 janv
1
A prod 175 tube verre bouchon paraffine
assez fort
22 janv
C — 500 – 40 2000 – 10
500 – 17 1000 – 15
C 2me face — 50 – 8
après 2 jours
à travers tube aluminium
B
100 – 13
100 – 14
100 – 18
Rien
bouchon paraffine

DEADLY SUNSHINE

THE HISTORY AND FATAL LEGACY OF RADIUM

DAVID I. HARVIE

TEMPUS

To Patrick Harvie MSP
All theory, dear friend, is grey,
But the golden tree of actual life springs ever green.
Goethe

Frontispiece: Pages from Marie Curie's laboratory notebook on the extraction of radium. (Wellcome Library, London)

First published 2005

Tempus Publishing Limited
The Mill, Brimscombe Port,
Stroud, Gloucestershire, GL5 2QG
www.tempus-publishing.com

British Library Cataloguing in Publication Data.
A catalogue record for this book is available from the British Library.

ISBN 0 7524 3395 4

Typesetting and origination by Tempus Publishing Limited
Printed and bound in Great Britain

CONTENTS

ACKNOWLEDGEMENTS

Since I first began to research this subject many years ago, I have gathered a huge number of correspondents who have, without exception, helped my understanding, and who have usually been willing to give me information that would not otherwise have been available. Often the best thing they did was to put me in contact with yet others, thus expanding my list of contacts exponentially. Most of them (for reasons of space) will have to remain unmentioned; some of them did indeed seek that anonymity; however, I am in debt to all of them.

I owe particular thanks to some. Mike McGarr, then Librarian at the Institution of Mining and Metallurgy in London (now at the Institute of Materials, Minerals and Mining) was an early enthusiast for what I seemed to be doing, and went out of his way to be of help. Iain MacArthur, of Orangeville, Ontario, a great nephew of John Stewart MacArthur, was a fine host and informant. I was lucky to benefit from the willing help of officials at the then Industrial Pollution Inspectorate of the Scottish Development Department in Edinburgh (now long defunct), who gave me access to all the historical papers relating to the Loch Lomond Radium Works (some of which are now at the

National Archives of Scotland). I am grateful to Penelope Bulloch, Librarian at Balliol College, Oxford University (and through her the Master and Fellows of Balliol), who allowed me access to the then uncatalogued papers of John MacArthur. As always, I acknowledge several other notable libraries and archives, especially the Public Record Office at Kew, the National Archives of Scotland, Glasgow University Library and the National Library of Scotland.

As a layman swimming at the historical end of such a seemingly daunting scientific pool, I have been greatly appreciative of the willingness of professional scientists to be friendly advisers, willing to try to prevent me from the worst excesses of enthusiasm and its inevitable companion – misinterpretation. Almost from the beginning, I have benefited from the guidance of Professor Murdoch Baxter, lately director of the IAEA Marine Radiation Laboratory in Monaco, and now again enjoying the rigours of Scottish weather. In more recent times I have had a fruitful and most genial correspondence with the Illinois epidemiologist Dr James Stebbings, whose personal involvement in studies relating to those who suffered radiation injury are very relevant to aspects of this book. John Large has been helpful in explaining contamination and other issues. To them I offer my sincere appreciation and thanks for their friendly interest and for trawling through manuscript drafts intended for a general, non-scientific readership. As always, my wife Rose has been a friendly but rigorous critic. Needless to say, any remaining failures in this book are entirely my responsibility.

In the search for illustrations, I have had exemplary help from Dr Paul Frame, of Oak Ridge Associated Universities in Tennessee, who is not only a most helpful man but has constructed a wonderful website relating to the history of radium products. I have had the helpful long-distance co-operation of staff at the US National Archives and Records Administration in Chicago, and at the Canadian National Archives in Ottawa. Gill

Pearce of Paignton in Devon has enabled me to use the superb photograph of the staff (including Marcel Pochon) at the South Terras Mine in Cornwall in the late 1920s. I am grateful to the Wellcome Trust Library and to the Special Collections Library of Duke University in Durham, North Carolina, and I also acknowledge the Society of Authors and the Bernard Shaw Estate for permission to quote from *The Doctor's Dilemma*.

INTRODUCTION

This book is naturally radioactive. Humans are naturally radioactive. The world we live in is naturally radioactive, and always has been. We all contain a natural mix of radioactive species, including radium. Whatever we eat, and every time we take a drink of water or have a cup of tea or sup a pint of real ale, we ingest naturally occurring radium, mostly transported to us in water. Sometimes radioactive species, such as the radium decay product radon, come at us as a gas from the very walls of our homes and workplaces, originating in the building materials and deep within the recesses of the Earth; other types of natural radiation emanate from the sun and from the outer reaches of the universe. There is also of course less natural, man-made radiation around us, as the result of man's activities (principally involving nuclear reactors and bombs). Radiation, that invisible 'monster' of the twentieth century, can of course kill, in rare situations of excess, and that singular fact has generated fear and loathing in degrees that should not be surprising: we have a uniquely sensitive nerve for the subject. But it is important not to assume that all radiation is somehow the evil end-result of a collusion between mad scientists, bad politicians and a crazed military; such an alliance may play a significant role, but that is not for this book.

Radium as found in nature is a radioactive element of the alkaline earth series of metals. It has the atomic number 88, the symbol Ra and an atomic weight of 226.0254. It is essentially the product of the radioactive disintegration of uranium, and is over a million times more radioactive per gram than its primary parent. Both are acquired in minute quantities from the same minerals, which were usually obtained from the same mines across the world. Radium emits alpha, beta and gamma rays, the latter two of which were used very successfully in radiotherapy; it has a half-life of 1,602 years and decays naturally to the radioactive gas radon. Pure radium metal, which has rarely been produced, is silvery-white in colour, blackening on exposure to the air; it is luminescent and reacts strongly with water. The term 'radium' as used in this book refers to the refined compounds radium bromide or radium chloride, which take the form of white crystals or powder.

Radiation consists of a variety of energetic particles, fast and slow moving, and electromagnetic radiation, of high and low energy; in everyday circumstances, such particles are undetectable. In large doses, radiation can kill within hours; in lower doses, illness as a result of damage to DNA and to blood characteristics can occur over a much longer period, often resulting in death. Radium can affect the body from an external source as the result of direct radiation by its highly penetrating gamma rays; it can also enter the body by inhalation and ingestion, when most damage is done by the less penetrating but potentially more dangerous alpha particles irradiating internal organs. Quantifying the risk from high doses is relatively easier than from low doses, and most scientists now accept the 'linear, no threshold' principle. In other words, risk is proportional to dose, and everyone should avoid even very low unnecessary doses from whatever source.

The story of radium is sufficiently interesting that there is no need to 'sex it up'. The radium contained in our real ale is identical to that which has generated lurid stories of death and disaster for the last century. Of course, there were problems and accidents coupled with a fair degree of ignorance and, at least to begin with, a considerable number of those miserable specimens of humanity inclined to take deceitful and dangerous advantage of their fellows. Undoubtedly, some of those who were heavily exposed in the very early years suffered hideously. But for the most part, radium was in fact quite difficult to acquire. Even among those who were exposed, most carried on with life largely unaffected. Once the problems of radiation exposure were recognised, in the late 1920s, diligent work was undertaken to ensure that regulation and control were established and these standards have been continuously revisited, modified and improved as the decades have progressed. The standards established for radium became the basis for all subsequent improvements in the wider nuclear environment.

The public generally allowed itself to be informed by an excitable and butterfly-like press. Sometimes the Fourth Estate was either uninformed or misinformed – both unsatisfactory – but behaved as if it knew better; the press was often superficial, and almost always on the lookout for the unique or sensational 'angle'. To begin with, they gave us the 'life-force mystery' of radium; then the 'fun' and later the 'horror'. There was no room for equivocation when discussing the potential for Armageddon; as is increasingly the case today, sides had to be taken and the only acceptable points of view were opposites: for or against, 'rays of life' or 'death rays'. Hollywood and literature fed on the fears and gave us the imagery. Boris Karloff, in the 1936 film *The Invisible Ray*, devised his 'radium ray projector', which was intended to perform good works but, inevitably, was turned to murderous use. H.G. Wells, in his prophetic novel *The World Set Free*, first used the term 'atomic bomb', and his cataclysmic world war offered both Armageddon and a universal Utopia. The prospects for the public achieving a balanced understanding were not high then and, with similar issues today, seem hardly better. Our perceptions today are assumed to be new and better informed, but the accompanying anxieties are just the same. There is the same impetus (and apparent willingness) to adopt or accept views from extreme ends of the range of arguments, and public trust in 'them' (scientists and politicians) is eroded in the face of real concerns about the relationship between technology and society.

One aspect of 'the radium craze' as it developed was that the public did become aware of the great benefit that radium eventually bestowed on medicine. Following close on the heels of the discovery of X-rays, radium was soon identified as having a huge potential in the successful treatment of a range of diseases, particularly cancer, from a distinguished lexicon that was at the forefront of public concern. Whereas X-rays offered a diagnostic benefit, it was soon realised that with radium the benefit was to be therapeutic, although it was some time before it was possible to quantify why some effects were clearly beneficial while others were damaging. There were certainly problems to solve: the question of how to use radium was by no means the most critical. Acquiring the mineral ores was difficult, and identifying chemical processes by which to refine the available minerals into the desired compounds was chemically intensive and time-consuming. The worst problem was cost – academic institutions could not afford the massive sums involved in acquiring tiny amounts of white powder for the purposes of research; hospitals, requiring greater amounts at regular intervals and for a wider range of procedures, found life even harder. Instead of co-operation, it transpired that the international status quo on radium supply would be driven by competition. Nevertheless, the contribution of radium to medicine can hardly be overstated, not least because it required a more progressive interdisciplinary approach than had hitherto been prevalent. Although its medical use began to reduce after about thirty years, in favour of safer, artificially produced isotopes, neutron therapy and more sophisticated nuclear medicine, the benefits conferred ensured that the place of radium in the history of medicine is assured.

While the main focus of this book is the history of radium – including, as the subtitle suggests, the more antisocial aspects – it will also relate in a later chapter how the object of the world's obsession changed from radium to uranium. The fact that both elements are so closely related, derive from the same minerals and are found in the same mines across the world is merely one of several justifications. Another is the fact that it may help to form the picture of how the discovery of radium led directly to the nuclear economy that now drives our civilisation.

This account is not written by a scientist, nor is it intended to appeal to a specialist readership. I hope it will appeal to that possibly mythical beast, 'the informed general reader' who has an appetite for both the unusual and the interesting. Certainly, there is a host of fascinating characters in this tale, and not a few disturbing episodes to be explored. With a subject that touches so many nerves and impinges on the important social and political issues of the day, it is also impossible to avoid some controversy. Nevertheless, I hope to have presented a useful and balanced historical account of events which still have a relevance to life today.

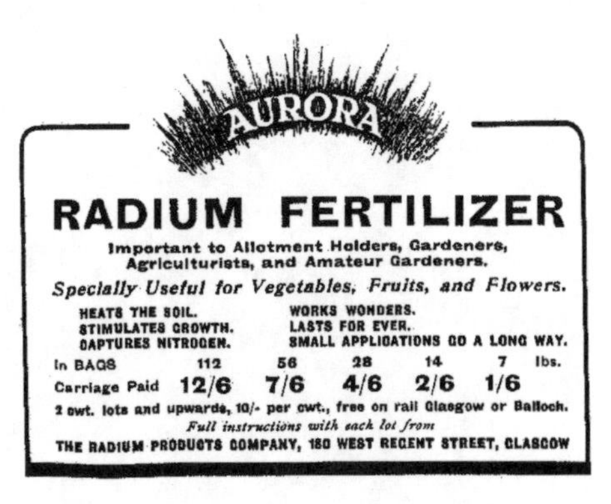

Advertisement for J.S. MacArthur's Aurora Radium Fertilizer. (See plate 13)

ONE

DOLLARS AND DWARVES

Erzgebirge, the ore mountains of Saxony and Bohemia – once part of the Austro-Hungarian Empire and now in the Czech Republic – make a wonderful backdrop for the hordes of international holidaymakers who converge on the area every year. They come to enjoy the landscape and the superb Gothic and Renaissance architecture: here stand some of the most spectacular castles, churches, monasteries and public buildings of Europe. Much sought-after local crafts include fine china and glassware, and delicate lace. Visitors can relax in vast palaces – once the playgrounds of European nobility – which now function as sophisticated spas and health resorts. At Karlovy Vary (originally Carlsbad), for example, where once Beethoven, Chopin, Schiller, Goethe and Wagner sought the rejuvenating air, tourists can relax surrounded by all the pampering demanded by the most self-indulgent of holidaymakers.

Bohemia's history has seen a good deal of political and social turbulence. Around the year 1200 Bohemia was an independent kingdom within the Holy Roman Empire, and enjoyed great wealth and success. But within a century, the Luxembourg dynasty had taken power. There were bitter disputes between Roman

Catholics and the Hussites (followers of the religious reformer Jan Hus, who was burnt at the stake in 1418). When the Luxembourg influence collapsed in 1437, the aristocracy gained power and the country was plunged into a kind of dictatorship in which the people were comprehensively deprived of rights and freedoms. In 1526, the Catholic Habsburg Archduke Ferdinand of Austria gained power. Religious disputes culminated in the Battle of the White Mountain in 1620, when the Habsburg Catholics defeated the Protestants; Czech nationalism was suppressed and German was imposed as the national language. There was popular protest against Habsburg domination in 1848, and gradually power passed from the nobility. In 1918, after the First World War, Czechoslovakia was formed and Bohemia became its principal industrial province. Following German invasion during the Second World War, there was a restoration of the status quo until 1993, when Czechoslovakia separated into Slovakia and the Czech Republic, the latter of which the ancient Bohemia is now a part.

The old royal town of Jáchymov lies at an altitude of 650m on the southern slopes of Mount Klínovec, highest peak of the Krušné Hory Mountains, 16km north of Karlovy Vary. This small town has been ravaged by both war, in the seventeenth century, and catastrophic fire, in the mid-nineteenth century. But the visitor can still see the Royal Mint, the early sixteenth-century Church of St Joachim and the late Gothic town hall. There are numerous natural springs, and winter sports have become a popular feature of the region. Opulent rooms can be taken at the beautiful neo-classical Radium Palace Hotel, where a vast range of therapeutic procedures is available for the cure and alleviation of all manner of ailments, both real and imaginary.

Over the centuries, Bohemia and Saxony – the heavily forested haunt of wolves and bears – were mined for a rich variety of minerals, principally copper, with some gold and silver. Much of this small-scale mining, centred on Schneeberg and Jáchymov, was often undertaken by itinerant or immigrant German miners, who gave the area a strong Germanic culture. In the earliest days, much of the silver mining was carried out from the cellars of houses, but in 1516 large silver deposits were discovered in the vicinity of Jáchymov (then known as Konradsgrün), and in the world's first 'rush' for precious metal, mines were established or redirected to win the valuable new discovery. In 1520, King Siegmund granted the town royal status and authority to mint 'Joachimsthalers', the German monetary unit; it was not until 350 years later that the Mark was adopted. The Thaler was the monetary progenitor which has given us by mutation, mispronunciation and misunderstanding, the dollar.[1]

In the early fifteenth century, copper miners in the Schneeberg area complained of a mysterious respiratory illness. These complaints spread throughout all the mining districts. The illness often proved fatal, and during the expansion of silver mining at Jáchymov, the frightened and superstitious mining communities began to devise irrational explanations for the feared mountain sickness or *Bergsucht* [2] (sometimes called *die Schneeberger Krankheit*, or 'the Schneeberg illness'). Although we now know that this part of Saxony has enhanced levels of background radiation due to geological conditions, it was clear to the miners, despite their lack of understanding of occupational poisoning, that the disease was caused by their employment.[3] Instead of being able to use rational enquiry, the communities sought dogged comfort in conjuring evil mountain dwarves, whose purpose was to punish the miners for the violation of their underground domain. However, Jáchymov had perhaps the best possible antidote to such folkloric reasoning.

Georg Bauer (1494–1555) – later known by the appellation Agricola – was appointed the town physician of Jáchymov in 1527. The settlement had been established eleven years earlier and was already a booming mining centre with a population of several thousand. Bauer was no ordinary small-town local doctor, though. He had studied philosophy, classics and medicine at Leipzig, and again in Bologna, Padua and Venice. He taught Latin and Greek and was the friend and collaborator of scholars such as Erasmus. Agricola seems not to have been enormously successful in the everyday practice of medicine, but he proved to have an extraordinary ability to comprehend and analyse, and to suggest vital social improvements based on informed observation. He made a detailed study of the mining industry, and of the working methods of the miners. He produced an immense, twelve-volume study of minerals and the mining industry, *De re metallica* ('On Metals'), first printed in Basle in 1555, just after his death. This outstanding work, illustrated by detailed woodcuts, appeared in many languages over the centuries and was regarded as being of world importance. In it he set out a humanitarian concern for the miners, who often worked in the most appalling conditions:

> It remains for me to speak of the ailments and accidents of miners, and of the methods by which they can guard against these, for we should always devote more care to maintaining our health, that we may freely perform our bodily functions, than to making profits. [4]

He said of the *Bergsucht* that:

> it eats away the lungs, and implants consumption in the body; hence in the mines of the Carpathian Mountains, women are found who have married seven husbands, all of whom this terrible consumption has carried off to a premature death.[5]

He devised many types of practical apparatus for the protection of miners underground, and one large-scale ventilation system for use in deep mines was operated by horses powering arrangements of bellows. Although Agricola had a sceptical response to the idea of supernatural activity, the presence of underground demons or gnomes (some benign, others evil) was such a commonly accepted belief that he was moved to accept the idea:

> In some of our mines, however, though in very few, there are other pernicious pests. These are demons of ferocious aspect, about which I have spoken in my book *De animatibus subterraneis* ['On Underground Spirits']. Demons of this kind are expelled and put to flight by prayer and fasting.[6]

It is an irony that the first man to write specifically about occupational disease was an alchemist and a contemporary of Agricola who did so after the study of miners in sixteenth-century Germany and Bohemia. *Von der Bergsucht oder Bergkranckheiten drey Bücher* ('Three Books on the Mountain Sickness') was the great work of a Swiss alchemist known for his drunken and disputatious eccentricity. But Theophrastus Phillipus Aureolus Bombastus von Hohenheim – thankfully more usually known as Paracelsus – was also a brilliant, cocksure individual who turned the arcane obsession of alchemy towards what became classical chemistry and the preparation of medicines for the cure and alleviation of disease. Despite his great ability and insight, however, Paracelsus was unable to discover the nature of the miners' *Bergsucht*.

The best practical efforts of the mining communities were directed towards fighting the invisible poison. At the suggestion of Agricola, they improved their lace-making techniques to provide miners with fine lace face-masks. While their efforts were completely in vain, a mine-master's wife, Barbara Uttman – 'the benefactress of Erzgebirge' – is acknowledged as the woman who introduced the tradition of extremely fine lace-making in 1561.[7] However, it was to be centuries before the cause of the hideous disease was known. The miners were unwittingly breathing radon, the gas which is a naturally occurring, colourless, odourless and tasteless 'radioactive daughter' of uranium and radium; estimates suggest that half the miners in the area died as a direct consequence of their occupation.

Alpha particles as emitted by radon are the least penetrating of radioactive emissions; but when radon and its daughters are ingested, 'stuck' to microscopic dust particles in the air, these massive, charged alpha particles cause great internal biological damage. Radon does most of its damage via its own daughters, polonium (^{218}Po) and bismuth (^{214}Bi), which both emit alpha particles – hence there are three alpha particles for each radioactive decay. The daughters, being particulate, readily attach to atmospheric dust, and are thus easily inhaled. (Another important radon daughter is the beta-emitter Lead ^{210}Pb.) Lung cancer was the long-term legacy bestowed on the miners by the mountain dwarves, although given the long incubation period for lung cancer, the immediate causes of death may strictly have been lung fibrosis or pneumonia.[8] Ironically, and prophetically, the cap-badge worn by the early miners of Schneeberg was identical to our modern international radiation warning symbol.[9]

The silver deposits which so suddenly brought prosperity and great national status to Jáchymov – or St Joachimsthal, as it became known – did not last long. Silver was found in greater deposits in the new American colonies and elsewhere, and in St Joachimsthal there was a lack of pumping and other equipment which would have allowed the mining of deeper levels.[10] By about 1560, silver mining was finished, but the ruling Habsburgs kept the mining industry ticking over. The miners turned their attention again to lesser minerals like bismuth, copper and cobalt. These were closely combined with other materials in mica-schist, making the miners' job more demanding. In particular, the miners were frustrated by large amounts of what they called an 'ore robber' – an unknown shiny black mineral which they called *Pechblende*, or 'bad-luck mineral', from which we derive the English 'pitchblende'. Where it existed, it seemed to repress the presence of more valuable ores. It was separated and laid aside in dumps with other unworkable material.[11]

Two centuries later, a piece of this strange material made its way to Berlin, where it was examined by an apothecary named Martin Heinrich Klaproth, who had a particular obsession with minerals. Klaproth, a follower of Lavoisier's theories, taught chemistry in the Berlin Artillery School and in 1810 became the first professor of chemistry at the University of Berlin. In 1789, having recently isolated zirconium from zircon, he announced that he had separated an oxide of the mysterious pitchblende (technically, the mineral uraninite, or uranium dioxide) from St Joachimsthal. Ignoring normal custom, he chose not to use his own name for the new element and instead named it after the last planet to have been discovered. 'Uranium' was named after Uranus, which had been discovered by Klaproth's countryman Sir William Herschel, the musician-turned-astronomer who lived in England. (While at this time the names 'uranium' and 'radium' were generally used to denote the dry, powdered salts of these elements, the names are more properly applied only to the metallic elements themselves, isolated in 1841 and 1910 respectively.[12])

The *Bergsucht* – the symptoms caused by the radon gas – was the result of the radioactive disintegrations within the pitchblende, not that Klaproth or anyone else knew that yet. The black 'ore robber' was still regarded as a useless problem. Paradoxically, after a few decades, the laboratory preparation of uranium salts revealed an interesting and finally useful property: they were capable of inducing remarkably bright colours in other materials. This was a largely forgotten property which had been exploited by Roman mosaic artists in Naples – examples of their work have been preserved at Pompeii by the eruption of Vesuvius in AD 79. It was soon realised in St Joachimsthal that they had great potential as colouring agents for glass and in glazes and paints for porcelain, pottery and leather.

In the 1840s, the derelict mines of St Joachimsthal were opened yet again. Significant amounts of pitchblende were already on the surface, having been dumped as rubbish, but new mining was also undertaken. Despite its frustrating prevalence during the silver-mining period, it was now nevertheless found to be quite difficult to mine in a commercially viable form. Nothing was as straightforward as simply shovelling the stuff up from the ground. It was found in veins varying from 6in to 2ft in width, but in complex association with galenite, fluorspar and various ores of silver, nickel, cobalt, bismuth, arsenic and copper – all of them in a variety of chemical configurations.[13] The pitchblende was treated at the mines to extract uranium. This was done by fusing the pitchblende with sodium sulphate. The uranium content was chemically converted to sodium uranate; after leaching out excess sodium sulphate the sodium uranate was dissolved from the insoluble residue by dilute sulphuric acid. Unknown to everyone at the time, the discarded residues contained the means of obtaining radium, via a complex chemical process that was still sixty years in the future.[14]

There are accounts from the 1840s of considerable use of uranium oxides in the colouring of glass and porcelain, and even one description of the use of uranium in colouring false teeth made from quartz.[15] However, the first commercial-scale use of uranium in glass production was probably undertaken at Park Glassworks at Springhill in Birmingham by the firm of Lloyd & Summerfield in 1850. Some claims suggest that the firm of Lobmeyr in Bohemia was first, but after Summerfield had travelled extensively in Europe, his company exhibited what are thought to have been the first examples of uranium glass at the Great Exhibition at the Crystal Palace in 1851.

Perhaps because of the expertise gained in Bohemia in mining and using cobalt in dye production, the government opened a pigment factory for the production of uranium dyes at St Joachimsthal in 1853. The main glass colouring agent produced a yellow colour with striking green fluorescence. Other colours available, depending on the particular uranium oxide and the manner in which it was used, were orange, red, green and black. The glass and porcelain producers of Europe soon vied with each other in the use of the new colours, always seeking to produce new effects and transitions. The prosperity of St Joachimsthal reached even greater heights than it had during its silver period. The techniques for making use of the uranium materials became quite complex, and uranium was sometimes dissolved with gold in aqua regia (one part concentrated nitric acid with three parts concentrated hydrochloric acid) to produce the most unique and startling colour effects*1. The success of uranium was such that the methods employed became a matter of great commercial secrecy and competition. In glassware, the once-despised 'ore robber' achieved a significance and a degree of concealment which foreshadowed its twentieth-century importance.

In the later nineteenth century, uranium was used in the form of uranyl nitrate in photography. The main exponent of this chemistry was Jacob Wothly, who patented his process in his native Germany and elsewhere, and favoured his process with the soubriquet 'Wothlytype'; it was well-received in some circles:

> We have lately seen some most beautiful specimens of Wothly or uranium pictures. They compare with the best silver prints, and would do honor to any photographer. In London, the Wothly collodion, also sensitized paper, which will keep in good condition for months, is now on sale; and at some of the photographic galleries negatives are taken, printed on paper by the Wothlytype process, and delivered to the sitter the same day.[16]

Around the same time, Ute and Navajo Indians in Colorado and Utah were using carnotite, another uranium mineral, for body paint and in the colouring of buckskins. Long before the element's radioactive properties were known, salts of uranium were used in a number of medical specialisms. As early as 1824 uranium was administered to animals with the intention of analysing its action on animal tissues. In 1860 homoeopathic physicians began to consider the value of uranium nitrate in treating human diabetes, on the homoeopathic principle of curing 'like with like'. Samuel West, a doctor at St Bartholomew's

Hospital in London, conducted experiments on eight diabetes patients in 1896, using a double chloride of quinine and uranium. Most of the patients showed significant improvement under West's treatment, regressing only in some cases when the uranium drug was removed because of intestinal irritation. Next came Ebenezer Duncan of Glasgow's Victoria Infirmary, who used doses ten times the strength of West's over extended periods. However, already-sceptical medical opinion soon turned against the procedure and it is thought that the last use of uranium in the treatment of diabetes was in 1917.

Other medical uses for uranium included the treatment of urinary incontinence in Buffalo, New York; the treatment of stomach ulcers; and in Britain, haemorrhage control and the treatment of consumptives in 1904. Apparently, Burroughs and Wellcome produced uranium salts in tablet form well into the twentieth century. Even more hopefully, a French medical concoction of 1930 combined uranyl nitrate, glycerol and red wine! [17] It seems fair to conclude that despite several decades of effort by a few individuals in different countries, the use of uranium salts in medicine was never fully accepted as efficacious.

In Britain, uranium ores were known to exist in Cornwall. This area of Britain has witnessed extensive mining activity from prehistoric times, and the prehistory is physically reflected in a large number of stone megaliths and circles. Although there were strong Roman and Saxon settlements here, there flourished a powerful Celtic culture which included a Cornish language and literature. In excess of 800 mine-sites have been identified in Devon and Cornwall alone, together with their dates, locations and principal outputs.[18] This list has been built from information which, given the passage of time and the lack of formal records, represents a considerable research effort. The principal ores mined were of tin, copper, manganese, silver and iron. The tin miners in particular became a powerful force in the community, and they were granted special crown privileges and a discrete legal jurisdiction based on the stannary courts (the name comes from the Latin for tin, *stannum*; the same source gives us the chemical symbol for tin, Sn). In the late twentieth century, deposits had to be sought and mined from greater and greater depths, at greater and greater cost. Simultaneously, cheaper foreign imports became available and when the world price of tin collapsed in the 1980s, the last tin mines in Cornwall closed. While it appears that the last vestiges of the Cornish mining industry are open-cast china clay workings near St Austell, attempts are being made at the time of writing to restart tin-mining activity.

There have been over seventy reported occurrences of uranium minerals in the Devon and Cornwall area, mostly only of historical, non-commercial, interest. As at St Joachimsthal, there was some mining of uranium for use in the European glassmaking industry. The first reference to uranium minerals dates from 1805, when oxide of uranium (thought to have been torbernite) was identified in mine dumps at Tincroft Mine near Camborne. In 1843, Wheal Trenwith near St Ives recorded 'pitchblende in great abundance

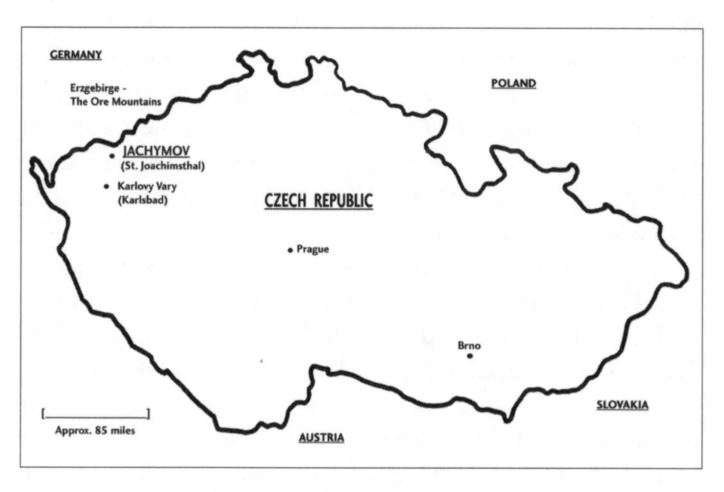

Map showing the location of St Joachimsthal.

among the copper ores'.[19] The largest pitchblende deposit was discovered at the Union Mine – an iron mine more commonly known as South Terras Mine – located by Tolgarrick Mill, between the villages of St Stephen and Grampound Road on the banks of the River Fal, 5 miles west of St Austell. South Terras worked an iron ore lode which intersected a tin lode. In 1873, a regular deposit of 'Green Jim' 4ft in width was discovered. This torbernite was found to exist in economically viable quantities. Previously, pitchblende – along with small amounts of silver, cobalt, nickel, bismuth and iron – was regarded as a spurious complication in the pre-eminent quest for copper and tin and was dumped as uneconomic waste.[20] Following the discovery of pitchblende at Trenwith and the despatch of samples for analysis, the report noted: 'Pitchblende being discovered, its nature and prejudice to the copper ores were explained to the workmen, by whom it has been of course since rejected.'[21]

From about 1830, a number of Cornish mines had exported small quantities of pitchblende or torbernite to Germany for use in the colouring of glass and pottery, and in the production of photographic chemicals. Between 1832 and 1865, for example, the Providence Mine near Trenwith produced 4,600 tons of tin ore, 9,720 tons of copper ore and less than one ton of pitchblende. The most productive British source of uranium ore was South Terras. In the twenty years from its reopening as a uranium mine in 1889 it exported 689 tons for processing at Brunswick in Germany. Two grades of ore were extracted – 36 per cent U_3O_8 and 3 per cent U_3O_8 – and, by mixing, two grades were sold for processing, 20 per cent and 11 per cent.[22] South Terras, like many of the other mines, passed through the hands of several owners during its varied life. The South Terras Tin Mining Co. Ltd (an odd title, given that the mine's principal output was iron) was wound up in 1888. In recognition of the new mineral, Uranium Mines Ltd became the new owning company the following year, with the substantial capital of £120,000 (the equivalent of perhaps

£5 million at today's values). However, within months the company was wound up, partly because it had no processing facilities of its own and was amassing large volumes of unprocessed pitchblende. Almost immediately, the mine was taken over by the Minerals Research Syndicate Ltd, and work continued.[23] The operation of some of Cornwall's mines was to change considerably when the world-shattering discoveries by Henri Becquerel and the Curies were announced early in the twentieth century.

The Bohemian *Bergsucht* was not reported in Cornwall, and here the dwarves wreaked their havoc, apparently unrecognised. As the nineteenth century drew to an end, events in Paris were about to explain the illness. But the entirely new and frightening future on which the world was about to embark – and in which St Joachimsthal was to play another part – would take much longer to explain.

TWO

LA RUE LHOMOND

Marya Salomea Skłodowska was born on 7 November 1867 in Freta Street in the old quarter of Warsaw, the fourth daughter and fifth child of two teachers whose family background and aspiration were those of Polish landed gentry. By 1867, however, that historical status in society was considerably faded, weakened by centuries of forlorn resistance to political and cultural torment. Just three years before Marya's birth, the Poles had seen the collapse of the January Uprising – another failed attempt to shake off the yoke of their Russian oppressors.[1]

The close, educated and quintessentially Polish family was humiliated when Marya's father Władysław lost money in a bad investment. He was then forced, for political reasons, into progressively more lowly academic posts. He blamed himself for years to come for his resulting inability to support his family in a sufficiently secure financial manner. But his passion for education survived, and Marya noted that: 'My father enjoyed any explanation he could give us about Nature and her ways. Unhappily, he had no laboratory and could not perform experiments.'[2] The family suffered grievous private wounds when Marya's eldest sister Zofia died of typhoid fever in 1876; worse, her mother died of tuberculosis two years later, when Marya was eleven years old, following a period of six years during which she had recuperated for long periods by 'taking the cure' at Innsbruck and Nice. Such tragedy was hardly unusual in Europe at the time, but Poland as a country was deeply repressed, with the native language banned and higher education for young women impossible. The family felt crushed.

Marya was a star pupil at school, but after being awarded a graduation gold medal, she suffered a collapse and was sent to join relatives in the country for a year to recuperate from what seems to have been a depressive illness. Her brother was studying medicine at the Tsar's University in Warsaw, but under Russian repression Marya and her elder sister Bronya had no chance of following that example. Instead, they courageously joined the 'Floating University' – a highly illegal night school organised by dissident Polish intellectuals and political revolutionaries. Teachers or pupils would have been imprisoned or deported to Siberia if discovered, and the classes were restricted to small numbers, in rotating clandestine locations, to avoid detection. With what seems like a supreme irony, the underground newspaper of this extraordinarily defiant institution was named *Pravda*. Despite the Russian oppression, Warsaw was a hotbed of patriotic, intellectual excitement. In the year 1889–90, there were a thousand women enrolled in the secret university and Marya and her sister were enthusiastic and self-confident dissidents.[3]

In the longer term, Bronya wanted to study medicine and the family joined forces to try to enable her to study in Paris. Marya, at the age of eighteen, took a job as a governess in order to help the situation financially. The *quid pro quo* was that with Bronya established in Paris, she would in turn help Marya to escape the constraints of life in Poland and to make her way to the freedom and academic possibilities of the French capital. During her time as a governess Marya was lonely and unhappy but regained her vitality by studying science and by establishing an illicit school for peasant children. She formed a relationship with Casimir

Zorawski, the son of one of her employers. Marya would have married him but his family regarded her as unsuitable for the son who would later become professor of mathematics at Warsaw Polytechnic. Five years after leaving Warsaw, Bronya was settled in Paris and married to a fellow Pole. The family's finances had improved sufficiently for Marya to follow her sister and, together with the offer of accommodation with Bronya and her husband and the attraction of a small Polish emigré community, everything at last directed Marya to Paris.

The Paris of 1891 was in that supposedly euphoric state of grace known as La Belle Epoque – that time of money, confidence and creativity when the well-to-do, at least, had never had it so good. The promise – following wars, siege and poverty – was that man would be master of his own fate, and that cultural and social levelling would lead to a more egalitarian future. When Marya arrived in the city, the extraordinary Gustave Eiffel's famous Tower had overcome the initial hostility of the artistic and chattering classes when it had opened two years earlier as the centrepiece of the Paris Exposition of 1889, celebrating the centenary of the French Revolution. The Tower, as the highest man-made structure in the world, had already achieved an iconic status that has never been matched.

Paris was a place of some political turmoil, however. A widespread belief developed that the Third Republic had failed the people of France. When the French company attempting to build a canal across the treacherous and disease-ridden isthmus of Panama faced financial catastrophe in 1891, members of the French parliament staged a conspiracy to prevent its collapse. Bankruptcy was nevertheless the result, together with financial disaster for many thousands of individual small investors. The aftermath also included the sentencing to imprisonment for fraud of Gustave Eiffel, who had designed the locks for the proposed canal. That mean-spirited sentence was later revoked on appeal, but the country was in chaos. In 1894 the infamous Dreyfus

Affair further appalled the nation, when a Jewish staff-officer in the army was the victim of a right-wing press conspiracy and was jailed for treason. Anarchist activity became rampant, climaxing with the assassination of the President of the Republic, Sadi Carnot. Artists, writers and scientists were all lumped together with 'modernists' and were assumed to be part of the cult of Reason and Science. In reply, the formidable novelist and critic Emile Zola wrote an influential 'Ode to Science' in 1895 and was an outspoken advocate of the Impressionists. He later staged his momentous attack on those corrupted officials who had pilloried Dreyfus with his open letter headed *J'accuse* in the journal *L'Aurore*. Every aspect of national culture was in uproar.

Marie Skłodowska, living in an attic and eating little, successfully studied for degrees in physics and mathematics which were completed in 1893 and 1894. She won a scholarship as the most outstanding Polish student, and was commissioned to undertake a study on the magnetic properties of various steels. In the spring of 1894, during the search for a suitable laboratory in which to complete that work, Marie was introduced by a mutual friend to Pierre Curie, who was laboratory head at the Municipal School of Industrial Physics and Chemistry in Paris. Marie later described her first sight of Pierre, standing on a balcony:

> He seemed to me very young, though he was at that time thirty-five years old. I was struck by the open expression of his face and by the slight suggestion of detachment in his whole attitude. His speech, rather slow and deliberate, his simplicity, and his smile, at once grave and youthful, inspired confidence.[4]

The ensuing liaison was to become a formidable meeting of minds, and would change the course of both science and world history. But the meeting inspired more than confidence. Both were both 'on the rebound', Marie from the friendship with Casimir Zorawski and Pierre following the death fifteen years earlier of a close woman friend; while neither was seeking, or expecting, anything other than a professional rapport, there was an immediate personal attraction. Curie, who was also studying aspects of magnetism, recognised in Marie an absolute determination, and courted her not with flowers but with signed copies of his research papers.[5] Pierre was awarded a doctorate for his work on magnetism in March 1895, was given a professorship, and four months later he and Marie were married in a civil ceremony.

The year of the Curies' marriage was to be a momentous one for physics. In Germany, Wilhelm Röntgen, Professor of Physics at Würzburg University, was working with his enclosed Crookes cathode-ray tube when he noted that on the other side of the room a glow appeared on a black screen coated with a fluorescent barium compound. He realised that unexpected and invisible penetrating rays were being produced by an unknown component of the cathode-ray tube. Further experiments – conducted with such preoccupation that his wife feared for his mental health – revealed that when he placed his hand between tube and screen an image of the bones in his hand appeared on the screen. He determined that photographic plates in a secure box became 'exposed' when placed in the same position. He later made a famous 'shadow picture' of the bones of his wife's hand, with her wedding-ring clearly in position. He discovered that, unlike light, these rays could neither be diffracted nor refracted, and that they were capable of passing through solid materials and still activating a fluorescent screen. He named the startling effect 'X-rays', and sent an account of his discovery to a local scientific society. In 1901 he received the first Nobel Prize for Physics for his discovery. A few weeks later the French physicist Henri Poincaré introduced Röntgen's paper to the Académie des Sciences in Paris, where Henri Becquerel, Professor of Physics at the Museum of Natural History in Paris, was inspired to make a study of the relationship between fluorescence and X-rays. Röntgen's initial interpretation was to produce a flood of would-be imitators of varying seriousness in the coming years; in 1896 alone there were 49 books and 1,044 papers.[6] For example, the French psychologist and sociologist Gustave Le Bon claimed his discovery of what he called 'black light'; and a few years later René Blondlot introduced his 'N-rays', named after the city of Nancy in which he worked. Blondlot's discovery was investigated in England and the USA but was eventually regarded less as a deception by Blondlot himself than as a case of mass self-deception by other physicists who for some years accepted a truth which did not exist.

After some initial lack of success, Becquerel decided to concentrate on a study of the uranium compound uranyl potassium sulphate, which had the known property of becoming fluorescent on exposure to sunlight; it would then fog a photographic plate covered in black paper. However, quite by accident, he discovered that the compound had the same effect 'without the action of the sun'. On a dull day he had decided to abandon the experiment, and placed the compound and a closed box of plates in a closed drawer together with a small copper cross. For some reason – was it accident or presentiment? – he decided to process one of the plates. The English physicist William Crookes, who happened to be visiting Becquerel's laboratory at the time, described the moment of the decisive discovery, when Becquerel developed the plate, expecting to find no fogging:

> To his astonishment, instead of a blank, as expected, the plate had darkened as strongly as if the uranium had been previously exposed to sunlight, the image of the copper cross shining out white against the black background.[7]

As surprising as the initial effect was the discovery that the phenomenon was long-lasting, as Becquerel himself described:

> The radiations of the uranium salts are emitted not only when the substances are exposed to light, but even when they have been kept

> in darkness, and for more than two months the same fragments of various salts kept secluded from any known exciting radiation have continued to emit new radiations almost without any appreciable decrease. From March 3rd to May 3rd the substances were shut up in an opaque box of cardboard. Since May 3rd they have been placed in a double box of lead, which remained in the dark chamber. Under these conditions the substances studied continue to emit active radiations.[8]

Marie Curie was immediately enthused by the prospect of studying the new phenomenon discovered by Becquerel. She obtained a grubby, poorly equipped laboratory at Pierre's School of Physics and Chemistry, although she had no funding or salary and had to rely on Pierre's financial and professional support. In the scientific community, the effects of uranium noted by Becquerel attracted little interest; the material was difficult to obtain, and most physicists were attracted to Röntgen's prior discovery of X-rays, which had the added fascination of spectacular images. Curie was influenced by the work of William Thomson, Lord Kelvin, at Glasgow University. He had known the Curies for some years and a few months after Becquerel's discovery had submitted a highly regarded paper on 'the electrification of air by uranium and its compounds' to the Royal Society of Edinburgh. Curie's biographer, Susan Quin, states that, 'When Marie Curie started her research, in the winter of 1897, she began where Lord Kelvin had left off'.[9]

There was considerable interest elsewhere in the early research into radioactivity, charged particles and rare gases. Frederick Soddy and Ernest Rutherford had been working at McGill University in Montréal. The latter – usually described as the founder of nuclear physics – discovered the different types of uranium radiations, and he and Soddy were both obsessed by the nature of this radioactivity. Together they formulated the Disintegration Theory, which for the first time suggested a transmutation process. Both men – later

Nobel Prize winners – returned to Britain. Rutherford worked with Hans Geiger at Manchester University, where at the end of 1918 he supervised the first splitting of the atom (provoking the *Sunday Express* headline 'THE ATOM SPLIT, BUT WORLD STILL SAFE'). Rutherford then became head of the Cavendish Laboratory at Cambridge. Soddy, the man who first defined the concept of radioisotopes, went briefly to the University of London before settling at the Chemistry Department of Glasgow University for ten years (followed by moves to Aberdeen and Oxford universities). In Glasgow, he worked with the two men who won the Nobel Prize for Chemistry and Physics in 1904; they were Sir William Ramsay (who identified the inert gases and is the only man to have discovered an entire periodic group of elements), and John William Strutt, Lord Rayleigh, who succeeded James Clerk Maxwell as Cavendish Professor at Cambridge. (Rutherford, Soddy and Ramsay were all mentioned several times by Marie Curie in her formal Nobel Chemistry Prize lecture in 1911.) Sir William Crookes was another towering figure in British science. He was the discoverer of thallium and inventor of the Crookes tube and the spinthariscope (and had been with Becquerel on that fateful day); he was a famous polymath and an enthusiastic exponent of the commercial application of scientific ideas.

Most of these internationally significant individuals were people of diverse interests in areas other than science. Ramsay was a good athlete, an accomplished musician and a gifted linguist. Crookes, while editing the weekly journal *Chemical News* for forty-five years and indulging in a truly incredible range of research interests, was also an impassioned spiritualist who had a life-long interest in telepathy and, as did Rayleigh, served as president of the Society for Psychical Research. (Pierre Curie was also fascinated by the study of 'psychic phenomena'.) Frederick Soddy found disenchantment in later life with his scientific work, and besides developing a wide range of other interests, became a somewhat frustrated political economist (one 1950 publication was entitled *Frederick Soddy calling all taxpayers*). Soddy was a complicated character who despaired that he had failed to persuade his fellow scientists – physicists in particular – to work for the real benefit of mankind.

Marie Curie did indeed embark on work in the area that Kelvin had been exploring, and began researching the activity of various uranium salts, trying to establish their propensity to ionise the surrounding air. After a few weeks she had established that the effect was dependent only on the amount of uranium present, and was unrelated to the particular chemical compound. The only other element she tried which had the same characteristics was thorium, and she deduced that the effect was related to the atom and not to chemical properties. For the first time, she used the term 'radioactivity' to describe this new effect. She compared the activities of naturally occurring uranium minerals (principally the 'ore robber' of the Bohemian copper miners, pitchblende, of which they had a meagre 100g). She discovered that some samples were as much as four times more active more expected, even though chemical analysis revealed only uranium and thorium to be present. She concluded that there must be an additional radioactive substance present, and reasoned that it was an unknown element. Curie made repeated chemical fractionations, and two were particularly radioactive: one containing mostly bismuth and the other barium. In July 1898 Marie and Pierre published a paper in which they described the hypothesis, and honoured Marie's birthplace:

> Two of us have shown that by purely chemical procedures it is possible to extract from pitchblende a strongly radioactive substance. This substance is related to bismuth by its analytical properties. We have expressed the opinion that perhaps the pitchblende contained a new element, for which we have proposed the name of polonium.[10]

Five months later they proposed a new element within the barium fraction, which she intended to call 'radium' from the Latin *radius*, or 'ray'. Marie described this fraction as containing 'a very large portion of barium; in spite of that the radioactivity is considerable. The radioactivity of radium then must be enormous.' Initial analysis suggested that the radium was almost a thousand times more radioactive than uranium; later, the figure was increased to a million. However, these conclusions were merely hypotheses; prodigious effort would be required before they could establish the identity and nature of the new elements. Chemists in particular were sceptical; they demanded characteristics such as physical appearance and atomic weight. The concept of radioactivity as a defining attribute was alien.

The Curies' small and ill-equipped laboratory was no longer big enough to accommodate their work, especially since Pierre had decided to join Marie in the demanding analyses. The Faculty of Medicine owned a nearby vacant shed, a disused dissecting room, off the rue Lhomond. This appalling space, with a leaking glass roof, broken plaster walls, an earthen floor covered with rough bitumen and one small cast-iron stove, was no better than their previous room, but was a little bigger. Eve Curie noted its main, ironic, advantage in her biography of her mother: 'It was so untempting, so miserable, that nobody thought of refusing them the use of it.'[11] The main need for more space was driven by their decision to try to acquire large amounts of pitchblende, although they had no clear idea whence this would be obtained.

The Curies wrote to Edmund Suess, who was president of the Academy of Sciences in Vienna, asking his assistance. Huge quantities of pitchblende would be needed, given that their original 100g had yielded only imperceptible traces of the new elements. Acquiring raw pitchblende itself would be prohibitively expensive. Perhaps a deal could be made to obtain residues from St Joachimsthal, which were lying in spoil-heaps in a pine forest. After treatment to extract the uranium used in ceramics colouring, tiny volumes of bismuth and barium would still be present in the crushed ore. The Austrian government eventually agreed to grant the Curies a ton of pitchblende residues, to be transported to Paris at their own cost; further supplies were to be obtained from the mine at whatever cost was mutually agreed. Her daughter Eve described the fervour with which Marie rushed into the street to greet the first rickety wagon-load:

> She cut the strings, undid the coarse sackcloth and plunged her two hands into the dull brown ore, still mixed with pine needles from Bohemia.[12]

This was the world's first radioactive waste, recognised for what it was, and Marie greeted it with such a passion. As they worked they would be unwittingly exposed to the invariable presence of radon gas, the alpha-emitting radon 'daughters' and the beta-emitter lead, ^{210}Pb, which produced severe bone damage. As well as physically handling radioactive waste, the Curies would be breathing its gaseous products in considerable concentration. The *Bergsucht* was to be their constant companion. Today, attitudes to such material are tightly bound up in scientific, medical, financial, commercial, social and political considerations of the most stringent nature. Marie and Pierre Curie were embarking on a project that would change the world. Marie in particular was both painstaking and stubborn; she would need to be. She thought at the outset that the radium proportion would be about one hundredth of the pitchblende; in fact, it was to be one millionth. They would need fifty tons of water and five tons of dangerous and corrosive chemicals to treat each ton of pitchblende residue; in all they treated eight tons of pitchblende over a period of four years, and produced one tenth of a gram of pure radium chloride.

Essentially, the separation of radium from other compounds is based on the chemical similarity of radium and barium salts. Both elements form carbonates and sulphates that are insoluble in water, whereas their chlorides and nitrates are soluble in water. The Curies' procedure involved cycles of pulverising and boiling up to 20kg of pitchblende at a time with acids and other chemicals in a succession of iron cauldrons; there followed relentless cycles of washing, boiling, and precipitation, with purification achieved by successive precipitation, dissolution and reprecipitation to separate the chemically similar radium and barium compounds. The complexity was described by Curie in 1903 in her doctoral thesis:

> The residue chiefly contains the sulphates of lead and calcium, silica, alumina and iron oxide. In addition nearly all the metals are found in greater or smaller amount (copper, bismuth, zinc, cobalt, manganese, nickel, vanadium, antimony, thallium, rare earths, niobium, tantalum, arsenic, barium, etc.). Radium is found in this mixture as sulphate, and is the least soluble sulphate in it. In order to dissolve it, it is necessary to remove the sulphuric acid as far as possible. To do this the residue is first treated with a boiling concentrated soda solution. The sulphuric acid combined with the lead, aluminium and calcium passes, for the most part, into solution as sulphate of sodium, which is removed by repeatedly washing with water. The alkaline solution removes at the same time lead, silicon and aluminium. The insoluble portion is attacked by ordinary hydrochloric acid. This operation completely disintegrates the material, and dissolves most of it. Polonium and actinium may be obtained from this solution; the former is precipitated by sulphuretted hydrogen, the latter is found in the hydrates precipitated by ammonia in the solution separated from the sulphides and oxidised. Radium remains in the insoluble portion. This portion is washed with water, and then treated with a boiling concentrated solution of carbonate of soda. This operation completes the transformation of the sulphates of barium and radium into carbonates. The material is then thoroughly washed with water, and then treated with dilute hydrochloric acid, quite free from sulphuric acid. The

solution contains radium as well as polonium and actinium. It is filtered and precipitated with sulphuric acid. In this way the crude sulphates of barium containing radium and calcium, of lead, and of iron and of a trace of actinium are obtained. The solution still contains a little actinium and polonium, which may be separated out as in the case of the first hydrochloric acid solution.

From one ton of residue 10 to 20kg of crude sulphates are obtained, the activity of which is from thirty to sixty times as great as that of metallic uranium. They must now be purified. For this purpose they are boiled with sodium carbonate and transformed into the chlorides. The solution is treated with sulphuretted hydrogen, which gives a small quantity of active sulphides containing polonium. The solution is filtered, oxidised by means of chlorine, and precipitated with pure ammonia. The precipitated hydrates and oxides are very active, and the activity is due to actinium. The filtered solution is precipitated with sodium carbonate. The precipitated carbonates of the alkaline earths are washed and converted into chlorides. These chlorides are evaporated to dryness, and washed with pure concentrated hydrochloric acid. Calcium chloride dissolves almost entirely, whilst the chloride of barium and radium remains insoluble. Thus, from one ton of the original material about 8kg of barium and radium chloride are obtained, of which the activity is about sixty times that of metallic uranium. The chloride is now ready for fractionation.[13]

Each stage in the seemingly endless series of fractional crystallisations produced progressive increases in the purity of the radium chloride obtained:

We found that by crystallising out the chloride of radioactive barium from a solution, we obtained crystals that were more radioactive, and consequently richer in radium, than the chloride that remained dissolved. It was only necessary to make repeated crystallisations to obtain finally a pure chloride of radium.[14]

There were no fume cupboards or other ventilation, and as each stage proceeded, the materials became more and more concentrated, radioactive, and dangerous; the final fractionations were especially likely to be ruined by the constant pollution of iron and coal dust in their inadequate laboratory. Every step had to be recorded in notebooks, and the results of each series of fractionations were measured and carefully noted for comparison with all the others. The Curies received some financial assistance from the Institute of France, and the company which had marketed Pierre's scientific instruments, the Central Chemical Products Co., was of considerable assistance in enabling them to modify their laboratory techniques for large-scale commercial production. They also had vital analytical help from their colleague André Debierne, who had organised their laboratory and experimentally identified the procedures most likely to be successful.

Marie took upon herself most of the physical work with the intensive chemical processes, while Pierre submitted the successions of precipitates to analytical testing and recording. Not only was it necessary to produce the desired substances, but they had to be subjected to the determination of atomic weights and the investigation of the chemical, physical, radioactive and physiological properties. The lack of money, together with the cold, wet and primitive working conditions, must have been dispiriting to say the least, and Marie always emphasised these shortcomings in her accounts of these years. However, she also said of this period:

> And yet it was in this miserable old shed that the best and happiest years of our life were spent, entirely consecrated to work. I sometimes passed the whole day stirring a mass in ebullition, with an iron rod nearly as big as myself. In the evening I was broken with fatigue. [15]

All of this was going on in circumstances in which Marie was regarded as nothing more than a doctoral candidate, studying for a degree at her own expense and in her own time. She was also female, at a time when and in ultra-conservative academic circles in which there was considerable misogyny. (To say 'nothing more than' belies the fact that when she was awarded her doctorate in June 1903, she was the first woman in France to achieve the honour.) The French Academy of Sciences was already aware of the Curies' activities; in 1898, Marie had been awarded a prize for her work on magnetism and her interest in radioactivity. The Academy had noted that, 'whatever the future of this scientific view, the research of Madame Curie deserves the encouragement of the Academy.'[16] For the time being, however, there was nothing 'official' about what they were doing or where they were doing it. What kept them going were their absorption in each other and their preoccupation in the work before them; and there were some special moments:

> We had an especial joy in observing that our products containing concentrated radium were all spontaneously luminous. My husband, who had hoped to see them show beautiful colourations, had to agree that this other unhoped-for characteristic gave him even greater satisfaction. [17]

This captivating luminous property was later to be a feature in both the public's hysterical appreciation of radium, and in Marie Curie's reluctance to recognise its danger. But for now, the Curies were happy to send samples of their precipitated salts to colleagues and correspondents across the world.[18] Sometimes in the evening, after they had gone home to eat at their two-storey rented flat on the Boulevard Kellermann opposite the Porte de Gentilly, they would walk back to the rue Lhomond in the dark for the simple pleasure of creeping quietly into their own dilapidated, gloomy laboratory to peer in awe at the rows of small crucibles glowing with a blue luminous radiance on the makeshift shelving. About this time, the respected German chemist Wilhelm Ostwald (who would win the Nobel Prize in 1909) arrived for a visit when there was no-one at the laboratory:

> At my earnest request, I was shown the laboratory where radium had been discovered shortly before. It was a cross between a stable and a potato shed, and if I had not seen the work-table and items of chemical apparatus, I would have thought that I was being played a practical joke. [19]

On 28 March 1902, the Curies were successful in amassing a tenth of a gram of radium chloride – sufficient and pure enough to enable their colleague Eugène Demarçay, an analytical specialist in the use of the spectroscope, to calculate its atomic weight as 225.93, thus ensuring the place of radium in the periodic table of elements.[20] They then investigated and set down its preliminary characteristics and properties (the pure metallic form of radium was not isolated until 1910). Marie described the properties of radium as 'extremely curious' and 'very complex'. It emitted radiation at least a million times more powerful than an equal quantity of uranium; the compounds were spontaneously luminous; it liberated heat spontaneously and continuously; it emitted a radioactive gas (radon) and its radiation fell into three different and distinct groups, each with different characteristics (which we now know as alpha, beta and gamma radiation).

There was, however, a mystery: where did this extraordinary amount of energy come from? The most basic and enduring principle of physics stated that matter and energy could neither be created nor destroyed – the atom was indivisible and energy was finite. The great Scottish physicist James Clerk Maxwell had stated that these assertions were 'the foundation stones of the material universe'. Radium appeared to break these sacrosanct rules by producing continuous, spontaneous internal energy. As early as 1898, Marie Curie had entertained the hugely radical idea that radioactivity might derive from the disintegration of atoms:

> Radioactivity is a property of the atom of radium; if, then, it is due to a transformation this transformation must take place in the atom itself. Consequently, from this point of view, the atom of radium would be in a process of evolution, and we should be forced to abandon the theory of the invariability of atoms, which is at the foundation of modern chemistry.[21]

The Curies shelved this bold thinking, at least for the time being, while the whole exciting argument became the property not only of practical physicists and chemists, but of philosophers. Correspondence increased dramatically between researchers across Europe, anxious to become part of the exciting new discussions and arguments. In England, Ramsay, Soddy and Rutherford supported Marie's theory by reporting, in 1903, that radium released small amounts of the gas helium, thus confirming the concept of transformation of atoms, and Soddy and Rutherford formally published their Disintegration Theory. Physics and chemistry were in turmoil.

But sexual politics was also bubbling; when *The Times* in London reported on radium for the first time, it did so by stating that Marie gave Pierre *invaluable assistance* in its isolation, but agreed that:

> it is obvious that M. Curie has introduced us to forces of a totally different order of magnitude. Heat sufficient to raise the mercury in the thermometer by 2.7 deg. is a different thing altogether. That effect must have a cause, for we are not to suppose that we have at last hit upon perpetual motion.[22]

In May, Pierre was invited to speak about their work on radium to the Royal Institution in London; as a woman, Marie could not be permitted to address this august body, but she was the star in the audience, and the event was a triumph. They brought a small vial of radium chloride as a gift for William Thomson,

Lord Kelvin, the seventy-nine-year-old patriarch of science in Britain, who showed his gift with pride to guests – he was the first person in Britain to possess radium.[23] Kelvin was nevertheless in philosophical disagreement with the Curies over the nature of radioactivity; he could not accept the concept of radium's internal source of energy and the subsequent implications for atomic theory.

The following month, Marie appeared in the Students' Hall at the Sorbonne in Paris in defence of her candidacy for a doctorate in physical science. This was based on her 'Researches on Radioactive Substances' and constituted a complete record of the work on polonium and radium. Surrounded by members of her family, who had travelled from Poland, the mood was one of happy celebration. By the end of the day, Marie Curie was the first woman in France to be awarded a doctorate. Later that year, Marie, Pierre and Henri Becquerel would jointly be awarded the Nobel Prize for Physics (after a thwarted attempt by some members of the French Academy to exclude Marie); in 1906, she was made the first woman professor at the Sorbonne. In 1911 Marie won the Nobel Prize for Chemistry, but again misogyny ruled when the double Nobel laureate was denied election to the French Academy. Despite narrow attitudes in French academic circles, she was nevertheless in prime position in the international scientific establishment, where she would remain for the rest of her life, despite the catastrophic loss of Pierre in a street accident in 1906, and subsequent controversy concerning her relationship with Paul Langevin, the prominent physicist. It is certain that both Marie and Pierre suffered as early as 1903 from radiation-induced ill-health, and Marie herself suffered a miscarriage that year. In December, she and Pierre were too ill to attend the Nobel Prize ceremony in Stockholm. Although she lived (in frail condition latterly, due to anaemia) until the age of sixty-seven, it has been estimated that she exposed herself during her work to radiation hundreds of times in excess of current dose limits. Today, her laboratory notebooks are secured, as they are dangerously radioactive. Her daughter Irene Joliot-Curie, who with her husband Frédéric were also outstanding workers in the field, died of leukaemia at the age of fifty-three.

As early as 1903, while the Curies were still attempting to characterise their new element, others, especially in Europe, were seeking out their own insights into the new substance, despite the expense and difficulty involved in its acquisition. In July, Frederick Soddy published the results of his investigations into the use of radium and thorium 'emanations' in the treatment of tuberculosis (or as it was called, even by such elevated scientists, 'consumption'):

> The immunity of these processes from external interference, the simple nature of the treatment proposed, the infinitesimal quantity of the active agents employed, the manner in which the emanations may be inhaled to do their work at the very seat of the disease, leaving behind in their place the excited activity to continue the work in a gentle manner after they have been exhaled, make out a strong case why the attention of medical men should be directed to these new weapons which physics and chemistry have placed at their disposal. [24]

In Dublin, work was underway using radium bromide to determine the potential for either improved germination, or for biological damage, in plant seedlings. This was to become a subject of much effort and some controversy within a few years (both in terms of horticulture and with reference to human biology), but the initial experiments with cress at Trinity College appeared to be inconclusive.[25] Professors Thomson, Rutherford and Dewar investigated the possibility that spa waters at Cambridge and Bath contained helium and 'a radioactive gas', and the presence of the radioactive gas – unidentified, but undoubtedly radon – was confirmed in a series of experiments conducted at the Blythswood

Laboratory near Renfrew.[26] The presence of radon in spa water would also become a lively issue in later years. These investigative activities were, in their various ways, the first faltering steps in the use of radium in academic research and in medicine. The phenomenon would be pioneering and controversial, and would lead to the establishment of a huge, worldwide and eventually largely military nuclear industry. Pierre Curie himself, when alerted to the medical possibilities, decided to kick-start things with an experiment on his own body.

THREE

TANTALISING HUMANITY AND ROUSING NOBLE EMOTIONS

Both Marie and Pierre Curie suffered physically from the conditions under which they worked; cold, dampness, hard physical effort, chemical fumes and long hours all took their toll. What they did not know was that they were unconsciously ingesting microscopic particles of radioactive materials and also breathing radon gas. It was to be many years before the damaging effects of radiation became slowly understood, and the methods that they employed took no account of the perils that were identified only later; if an accident occurred in the laboratory and a spillage resulted, someone would simply wipe up the mess with a rag. However, Henri Becquerel and the Curies did know that radium affected the skin of the fingers and hands in ways that were similar to the effects of X-rays. Pierre Curie suffered very badly from severe pain in the bones of his legs and back which limited his movement and led him to fear a life of severely limited mobility. He was being treated for 'rheumatism' with strychnine and a diet devoid of both red wine and red meat, and once confessed that he would not trust himself in a room with a kilo of pure radium, as it would burn all the skin off his body, destroy his eyesight and probably kill him.[1] Curiously, he made no intellectual connection between his work and his physical symptoms, despite the fact that laboratory animals that had been subjected to a high level of radon within a confined space all died in a few hours. He nevertheless acknowledged in a formal paper on *their* condition 'the reality of a toxic action from radium emanations introduced into the respiratory system'. [2] Ernest Rutherford was also unable to recognise the potential hazard. When a colleague working with radioactive material discharged an electroscope by merely exhaling in its vicinity, Rutherford was delighted to have this confirmation of the existence of airborne emanation, without apparently considering the biological implications.[3]

The possibility that radium might have therapeutic benefits similar to those of X-rays was intriguing, and a number of the Curies' colleagues encouraged them in that thinking. However, they also knew that radium salts in close skin contact produced discolouring, and sometimes blisters similar to burns, and that if the exposure was extended, ulceration occurred.

> In one experiment, M. Curie caused a relatively weak radioactive product to act upon his arm for ten hours. The redness appeared immediately, and later a wound was caused which took four months to heal. [4]

The Curies gave small amounts of radium to Dr Henri Danlos and his colleague Dr Block at the Hôpital St Louis in Paris, who reported their findings on its use on patients with skin complaints. In June 1901, Pierre Curie and Henri Becquerel jointly published a paper entitled 'The Physiological Effects of Rays' and this provoked widespread studies on the effects of radium, borrowed in tiny quantities from the Curies, on both healthy and unhealthy tissue. Radiotherapy (or 'Curietherapy' as it became known in France) was thus identified as a new branch of medicine even before Marie Curie published her thesis on radium.

As the new century began, scientific interest became intense in the new and exciting subject that encompassed both physics and chemistry and which offered a new intellectual approach to 'natural philosophy'. In the first ten years of the new century, science sought to understand what radioactivity actually was. In Britain, as elsewhere, years were to pass before any means of large-scale radium production could be contemplated or brought to fruition. A great deal of purely scientific investigation took place within university laboratories in attempts to achieve an understanding of radioactivity and its implications. With few native ore resources, much of this experimental activity took place as the result of British researchers being gifted small samples of radium materials from the university communities of Europe, although there were occasional tensions and disagreements between institutions over loan and sharing arrangements.

In the USA, the Cornish metallurgist Richard Pearce was involved in the first discovery of pitchblende in that country. As early as 1859 he had found the mineral in Cornwall, and after being contracted by the Rochdale Mining Co. of London to investigate gold mines in Gilpin County, Colorado, he found pitchblende there too at Leavenworth Gulch, just west of the town of Central City, forty miles from Denver. He later described his find:

> In the summer of 1871 the writer's attention was directed to a group of mines in Leavenworth Gulch, owned by the Rochdale Mining Co., and in the course of his examination of one of the several claims belonging to this company, he found on the dump of the Wood claim a heavy black mineral which proved to be uraninite coated with a beautiful canary yellow material, uranium vitriol, a basic sulphate of uranium formed from the oxide by lengthened exposure on the surface.[5]

As everywhere, the pitchblende had been regarded as an expensive nuisance, and in the search for gold and silver, it was consigned to the mine dumps in some quantity. Pearce shocked everyone by telling them that it was worth £400 ($2,000) per ton in England. The ore was found to be present at a number of other small mines in the area and until the discovery of radium, it was processed abroad for use in the glass and ceramics industries. When its radium-bearing property was discovered, the interest in finding viable pitchblende supplies intensified. However, it was found that deposits were highly localised, and enthusiasm waned after the French chemist Charles Poulot discovered another mineral that, when analysed in Paris, showed high uranium content. This was a bright yellow uranium-vanadium mineral which, when its value was recognised, was named Carnotite, after Adolphe Carnot, an eminent French chemical engineer. This was potentially a much more commercially viable mineral than the pitchblende of Gilpin County, and interest immediately switched to acquiring it in quantity. Deposits were found in south-western Colorado, in the Paradox Valley in Montrose County, and in eastern Utah at Richardson, near Cisco, just over the Colorado state line, 180 miles south-east of Salt Lake City. In 1900 Stephen Lockwood and his colleagues in the Welsh-Lofftus and Rare Metals Co. obtained the advice of Pierre Curie, and subsequently chemical reduction plants were established in 1903 at Lackawanna in the great steel-producing area around Buffalo and Niagara Falls in New York State. These initially produced uranium oxide and iron vanadate for use in the metallurgical and steel-producing industries, until the opportunity came to produce radium using improved processes.[6] The Lackawanna plant was the first in the USA to produce radium commercially, but the project did not live up to initial promise. Chemical costs proved extremely high, and by 1908 disappointingly low quantities and grades of carnotite led to production being terminated.[7] Today, the steel industry has also gone, and the area that was once characterised as 'Steel Town USA' is now part of the American rust belt.

While geologists began the search for likely radium ores, chemists were experimenting with simpler or less time- and chemistry-intensive extraction processes. For years, there was even some confusion about what the term 'radium' actually meant. It was 1915 before the US Geological Survey published the best general description in its summary of national mineral resources:

> The Geological Survey has received numerous enquiries about the appearance of radium, many of which seem to have been prompted by carelessly written stories appearing in periodicals of the poorer class; these stories tell of remarkable finds of nearly pure radium.
>
> Radium is a metal, and is described as having a white metallic luster. It has been isolated only once or twice, and few people have seen it. Radium is ordinarily obtained from its ores in the form of hydrous sulphate, chloride, or bromide, and it is in the form of these salts that it is usually sold and used. These are all white or nearly white substances, whose appearance is no more remarkable than that of common salt or baking powder. Radium is found in nature in such exceedingly small quantities that it is never visible even when the material is examined with a microscope. Ordinarily radium ore carries only a small fraction of a grain per ton of material, and radium will never be found in large masses because it is formed by the decay of uranium, a process which is wonderfully slow, and radium itself decays and changes to other elements so rapidly that it is impossible for it to accumulate naturally in visible masses. Radium and radium minerals are not generally luminescent. Tubes containing radium glow from impurities present which the radiations from the radium cause to give light. [8]

By 1903, the science journal *Nature* was regularly reporting new discoveries and investigations. Prof. Sir James Dewar and R.J. Strutt (the son of Lord Rayleigh) had discovered helium in the hot springs at Bath and Buxton, and further studies of radon in tap-water were being carried out by them and by

Prof. J.J. Thomson and others, with analyses being conducted at the Blythswood Laboratory in Renfrew.[9] Dewar also conducted positive tests for radon in Birmingham, Ipswich and Ely, and at a number of sites in Cambridge – including Trinity College Well, the Star Brewery, Girton College and Trinity Hall Cricket Ground.[10] The helium discovery at Bath soon led to talk of radon and radium, as Strutt informed the Baths Committee:

> I think there can be little doubt that the Helium of Bath owes its origin to large quantities of Radium at a great depth below the earth's surface. A little of this Radium is carried up by the rush of hot water and is found in the deposit. My experiments promise further interesting developments, which I shall have much pleasure in bringing to the notice of the Committee in due course. [11]

Frederick Soddy was using radium and thorium in the treatment of consumption, and others were testing the effects on plants. In Vienna, Professor Gussenbauer cautiously reported the complete disappearance of a cancerous tumour, but 'the early publication of these details in the public press before there has been time to test the method effectually is much to be deprecated'.[12] Alexander Graham Bell, the Scots-born inventor of the telephone and gramophone and inspired teacher of deaf-mutes, was extremely interested in the medical uses of radium. In 1903 he wrote to the journals *American Medicine* and *Nature* from his Cape Breton home at Baddeck, Nova Scotia, describing his efforts to encourage doctors in Washington, with whom he had been in correspondence, to take up the cause of radium treatment (his proposals were already being adopted in France):

> I understand from you that the Roentgen Rays, and the rays emitted by radium, have been found to have a marked curative effect upon external cancers, but that the effects upon deep-seated

> cancers have not thus far proved satisfactory. It has occurred to me that one reason for the unsatisfactory nature of these latter experiments arises from the fact that the rays have been applied externally, thus having to pass through healthy tissues of various depths in order to reach the cancerous matter. The Crookes' Tube, from which the Roentgen rays are emitted, is of course too bulky to be admitted into the middle of a mass of cancer, but there is no reason why a tiny fragment of radium sealed up in a fine glass tube should not be inserted into the very heart of the cancer, thus acting directly upon the diseased material. Would it not be worth while making experiments along this line? [13]

In the main cities, Soddy, Ramsay, Strutt and others were giving lectures in universities, hospitals and municipal chambers, often accompanied by demonstrations of the effects of radium. In the atmosphere of what seemed like the discovery of the miracle treatment of the new century, nothing could be explained with certainty and there was a real, if confused, sense of unease and awe at these events. Even within the strictly medical context, early experiments were too often reported as dramatic rather than conclusive. In December 1903, Marie Curie published her method for the preparation of radium chloride; the Curies never patented their work, which remained open to science the world over. They had discussed the issue of patenting, but Marie had insisted: 'It is impossible. It would be contrary to the scientific spirit.'[14] During all the excitement, there was also controversy. Lord Kelvin was involved in 1906, the year before his death, in a raging squabble in the press with Soddy and Ramsay over the suggestion of the 'transmutation' of radium to helium.[15] This furore was joined by Marie Curie in the Parisian newspaper *Le Figaro*, agreeing with Ramsay, Rutherford and Soddy and saying that Kelvin's reported views must have been misinterpreted: 'the discussion may have turned upon words rather than upon concrete ideas'.

In 1907, the work being done on radium in France was moving away from the unique preserve of the academic research chemist. A new if crude technology was about to appear. A large radium plant was built at Nogent-sur-Marne, about 50 miles north of Dijon. The new plant (which photographs depict as distinctly industrial in character) used pitchblende from Bohemia, Canada and Colorado and carnotite from Utah, Portugal and Madagascar, and followed the process described by the Curies, who had their own laboratory at the site:

> The Manufactory lately installed at Nogent-sur-Marne bears but little resemblance to ordinary establishments; whole carloads of divers minerals are there treated for an ultimate product consisting of a few minute particles of radium salts.
>
> Is not pure radium bromide, for instance, a new and most remarkable philosopher's stone, seeing that a single kilogramme of this precious substance would be worth 400,000,000 francs at the present time? It is well-known that radium has not yet been isolated in a metallic state, but exists only in the form of salts (chlorides, bromides or sulphates) possessed of a greater or less degree of activity, secured by stopping the operations of manufacture at certain determined points. The activity (taking uranium as unity) reaches 50 to 60: the first fractionings beyond this raise it to 1,000 and the final ones carry it to 2,000,000. [16]

In 1908, another controversy raged in the scientific press. Ramsay had claimed at a British Association meeting that he thought he had succeeded in transmuting copper into lithium by the action of radium. However it was generally concluded that Ramsay – whose reputation in detecting atmospheric gases was unsurpassed – was over-reaching himself in the new field of radioactivity. Marie Curie repeated Ramsay's experiment without success, and it was suggested that Ramsay had contaminated his experiments, giving him false results. An anonymous 'finger-wagging' letter appeared in *The Times*, pleading for more self-control on the part of the scientific community:

> It may be said, without fear of contradiction, that the powers of radium have been vastly over-rated. Itself a most mysterious substance, we have yet to know the exact nature of the changes which it undergoes. Speculation on a most slender basis of fact has played far too great a part in the enquiries hitherto carried out with it.[17]

Demands for caution and discretion became insistent: caution in using radium, and discretion in talking about it. Ernest Rutherford allegedly warned Soddy in 1901, 'Don't call it transmutation. They'll have our heads off as alchemists!' Soddy and Rutherford had contaminated their laboratory with radium, for a time giving rise to the serious scientific concern that radioactivity might be contagious. They soon knew that this was wrong, but the idea caught hold in the public's imagination that a single radioactive atom would inevitably make adjacent atoms radioactive, initiating an unstoppable global chain reaction. The 'End of The World' delusion was one which retained public currency for many years and popular perceptions of the new radioactivity were promoted by two camps. The 'Rays of Life' faction celebrated the seemingly magical benefits; and the 'Death Ray' brigade characterised the mystery 'energies' as inevitably malevolent and poured scorn and doom on the evil science.

Popular imagination was fired by sensational publicity; when the New York Museum of Natural History displayed a speck of radium in 1903, the largest crowd ever to enter the building fought for a close look at what amounted to a few grains of powder. As with X-rays a few years earlier, radium was to suffer the twin portrayals of Good and Bad; the 'all-seeing ray' became a literal image for the gullible. Later, when it became possible to buy 'X-ray Spectacles', or 'Radium Spectacles', which claimed to enable the wearer to 'see through things', competing charlatans offered special underwear which prevented the magic spectacles fulfilling their evil promise. In the hysteria, the legislature in the US state of New Jersey considered introducing a bill banning the use of X-rays in opera glasses.[18] The press talked up the fun, the mystery and the unfathomable potential of the new 'magic'. In *Punch*, the year's review of 1904 depicted Harlequin Science revealing the Good Fairy Radium and her magical powers. There were cartoons on radium lectures, and on radium in the waters of the Pump Room at Bath; one shows a little girl asking the grocer for 'a pennyworth of this radium everyone's talking about'; and a young couple dream of being married and cooking by radium ('But we won't have to wait for that to get married?' queries the sceptical young woman). Mr Punch summed up the year's events by poetically quoting some popular fallacies about what was nevertheless 'an exceedingly dangerous substance':

> Radium, very expensive, the source of perpetual motion,
> Take but a pinch of the same, you'll find it according to experts
> Equal for luminous ends to a couple of millions of candles
> Equal for heat to a furnace of heaven knows how many horse-power.[19]

Large advertisements began to appear in the press, like that in the autumn of 1908 by the Cardinal Manufacturing Co. of Moorgate in London. This headlined the 'New Radium Discovery' and suggested its future as a cure for cancer, following the receipt of 'an interesting cablegram from St Petersburg'. The company was producing a 'radium salve' for use on the skin:

> There is absolutely no danger with this. Just a tingling of the skin is felt. The tingling is most pleasant. The men in the London warehouse state that they use it when they have headaches or toothaches, and that the 'Salve' almost instantly relieves the pain.[20]

Despite the unconvincing approval of London warehousemen, and the confusion being sown by the press, there were several promising proposals for the use of radium in medicine. Tiny amounts of radium salts could be sealed in small glass or platinum tubes, which would then be placed in close contact with the affected tissue for a number of hours, as the journal *Nature* reported:

> Radium has usually been employed in the form of crystals of the bromide. These crystals are contained either in a sealed glass tube or in a button with a covering of thin glass, aluminium or mica. Recently the crystals have been spread in a thin layer upon a flat surface and covered with a layer of varnish. Such buttons and spread preparations are suitable for application to the surface of the body. The glass tubes may be inserted into the interior of tumours, or into orifices of the body. [21]

The cost of radium was an over-riding factor, however. Many physicians were trying the 'emanation' method, which only required one small quantity of radium, which was repeatedly used to produce 'radium emanation' or radon gas. The radon could be used directly, by inhalation, or by infusing water or other materials with its therapeutic benefits. Since the radium salts would be capable of producing radon by natural decay for perhaps two thousand years, the costs were greatly reduced. Another benefit, not then fully realised, was that radon had a 'half-life' of a few days, and was therefore inherently a safer prospect than using pure radium salts:

> As it is risky to send a patient away with a valuable tube of radium crystals within his body, sealed glass tubes of radium emanation have recently been used. They are enclosed in lead tubing one millimetre in thickness. Such tubes, of about 10 milligramme strength, may be placed in contact with a cancerous growth and allowed

> to decay in situ. At the end of a fortnight they may be removed, as being then too weak to be of further use. [22]

In various parts of the world, the search was underway for uranium ores by 1910. In France, where the radium business was well developed, autunite was being extracted at Autun and Limoges, for processing at Nogent-sur-Marne along with the pitchblende from Bohemia and Portuguese carnotite. A 'Banque de Radium' was planned by M. Henri Farjas, who had patented several methods of impregnating medical materials with radioactivity. Farjas was also the editor of a monthly journal, *Le Radium*, which had been published in Paris since January 1904 and which was a forum for all matters relating to the production, use and ethics of using the new element. The journal also printed fascinating photographs illustrating what appear to us as grossly unsophisticated methods of production. Marie Curie herself complained of having to stir a boiling cauldron with an iron ladle as big as herself.

In Sweden, a company was formed in 1909 to produce radium from a coal-like mineral named 'kolm'. It had to be burned to a fine ash, which contained about 2.5 per cent uranium, which itself was expected to contain one millionth part radium. It was hoped that one ton of kolm would produce 5mg of radium sulphate.[23] In 1909, the price of radium chloride – if you found someone able to supply it – was about £20,000 per gram (over £750,000 at today's values). The scientific and medical communities were concerned to find both sources of the appropriate ores, and the expertise to undertake the complex and time-consuming refining processes. The Austrian government, recognising the potential value of the pitchblende in Bohemia, banned its export, making the problem of supply even more difficult. The meagre British output from the mine at Grampound in Cornwall, which was producing small amounts of pitchblende, was especially costly as it had to be exported to France or Germany for processing.

Information from Eastern Europe was hard to come by, but the *Mining Journal* gave some indications in 1911. In the huge Russian territories, radioactive minerals and springs were being developed. In the Finnish territories of Veaborg and Uleaborg, monazite had been found by 1910 and there were hopes of considerable thorium–uranium veins. Torbernite existed by Lake Onega in North Russia, and there were expectations of substantial deposits of uranium and thorium minerals in the Urals. Eastern Siberia, Ferghana in Central Asia, and the Caucasus were all thought likely to produce carnotite and pitchblende.[24] In Australia too, a number of searches proved successful; the Baker's Creek Gold Mining Co. began mining uranium ores 9 miles from Broken Hill in 1910; pilburite was exploited in Western Australia and pitchblende in South Australia. In December 1912 the first radium bromide produced outside Europe was revealed in Sydney; 400mg had been purified to a high degree from Australian ores, and there were hopes of producing 40mg per week.[25]

In London, the Goldsmiths' Co. gave £1,000 to the Royal Society to purchase radium for research. But there was the problem of where to obtain it. In the end, there was a degree of political manoeuvring involving the reluctant Austrian government, the Royal Society and the Prince of Wales. Half a ton of residues was eventually sent from Austria to Paris, where it was partly refined, before being delivered for further fractionation at the Royal Institution. A breakthrough came when the National Minerals Corporation, through its subsidiary St Ives Consolidated Mines Ltd, discovered pitchblende at its Wheal Trenwith copper mine in Cornwall. Sir William Ramsay was engaged to extract radium bromide on an experimental basis, and when this proved successful, it was decided in June 1909 to form the British Radium Corporation, to produce radium on a commercial scale for the first time in Britain. The general manager of the mine at Grampound, T.H. North, made rather a lot of petulant noise in the press pointing out that the St Ives mine did not deserve all the publicity it was receiving, since the Grampound mine had been in operation first: 'This company will proceed with radium manufacture in its own quiet way without any flourish of trumpets.' [26]

About the same time, plans were being laid for the formation of a national Radium Institute to secure supplies of radium and carry out medical research. Sir William Ramsay, Sir Frederick Treves and R.J. Strutt joined a committee of the great and the good which was to oversee the establishment of the new body. The Institute – in Ridinghouse Street, near Portland Place in London – was to be largely funded by two philanthropists, Sir Ernest Cassel, the fabulously wealthy German-born financier, and Lord Iveagh, a member of the Guinness brewing family. In addition to cash, they planned to donate 7.5g of radium bromide, to be refined at Brunswick in Germany. Meanwhile, *The Lancet* reported success in the use of radium in various simple forms of treatment, such as 'a case of Spring Catarrh' and 'a case of a Lad who suffered from Warts on the Hand', who had been treated and cured after one application for 45 minutes of one centigram of radium bromide.[27] The medical world was still very uncertain about the radium issue – largely because not enough of it was available to enable proper evaluation to take place. The journal *Nature* sought to calm the hysteria:

> The supply of radium available for treatment of disease is still so limited that the therapeutic usefulness of this agent has not yet been fully determined. No sooner were indications noted of a prospect of relieving cancer by the use of radium than all the radium obtainable was devoted to this purpose; consequently, its action in less serious ailments is still almost unknown. This branch of therapeutics is still in its infancy, and it would be a mistake either to raise delusive hopes because some cancerous growths have been made to disappear under its use or to declare it useless because disappointments are common. [28]

Despite such strictures, the 'radium industry' was well underway in Britain. On 18 October 1909, Lady Ramsay took home with her an inscribed solid silver and ivory trowel following her ceremonial laying of the foundation stone of the British Radium Corporation's new radium works at Thomas Street, Limehouse, in London's East End. The works were at Baltic Wharf, on the Limehouse Cut, between a biscuit factory and a gold and silver refining works.[29] Sir William Ramsay, the company's chief chemist, proposed to use a process devised by himself and the company's manager, Norman Whitehouse, which, it was claimed, would reduce the refining time from nine months to eight weeks. Whitehouse was assisted by one of Ramsay's laboratory assistants, Donald Matthews, and by Ramsay's son, W.G. Ramsay; Ramsay senior himself often referred to the Limehouse plant as 'Willie's factory'.[30] One of the Corporation's directors extolled its virtues at the opening ceremony:

> Today pitchblende occupies the position of being one of the rarest minerals known, the comparative value of the crude ore being greatly in excess of the gold quartz of Johannesburg, or the blue ground of the diamond mines. Consequently the directors of this corporation thought it right, under the advice of their chief consulting chemist, Sir William Ramsay, to take steps to retain this important industry in this country. [31]

At the annual general meeting of St Ives Consolidated Mines it was announced that Ramsay had been testing the water in the Trenwith springs and he had declared them 'ranking with the most radioactive known'. The company, overcome with enthusiasm, decided to set up a subsidiary:

> Negotiations are in progress with persons of repute and experience in the hotel and spa business with a view to the establishment of high-class hotels, hydros and curative bathing establishments according to the most up-to-date methods. [32]

The enthusiasm was infectious, if other reports were anything to go by:

> Enumerating a few of the benefits which either drinking of or bathing in the waters might be expected to confer, the Chairman incidentally referred to their 'rejuvenating effect' and then every man in the room over 50 years of age chuckled with joy as he anticipated the day when he would play Faust to the Trenwith Mephistopheles. [33]

In March 1910, reports were received from Austria about the government's increased attempts to keep the radium business under firm control. Following their earlier decision to embargo exports, they were now intent on keeping a tight rein on the burgeoning 'health spa' industry which was expanding in and around St Joachimsthal. The highly radioactive waters at the spa were enough to provide over 500 baths per day:

> The bathing is now under the control of a public officer of health, and the efforts of the medical profession have averted the danger of this important health resort coming under the influence of unscrupulous speculators. Whilst last year the arrangements for bathing in the radioactive waters were rather primitive, this year first-class accommodation for patients has been supplied by the corporation, and health-seekers can enjoy their benefit all the year round. [34]

The 'Radiumkurhaus' also sold radium bromide directly to its customers. It was supplied, in varying strengths, in small brass and lead capsules, hallmarked to assure its quality. Each capsule contained one milligram, priced at 400 kronen (sixteen guineas, or about £675 at today's prices). A lead plate on the top could be swung aside, leaving the radium covered only by a thin mica sheet. Quite what the purchasers were supposed to do with these dangerous capsules is anybody's guess; at any rate it was done entirely unsupervised.

In 1910 came the establishment in London of Radium Ltd. This company was formed to enable its German sister firm, Radium Heil Gesellschaft of Berlin, to export radioactive water to be added to the already radioactive water at the Hydropathic Hotel in the spa town of Buxton in Derbyshire. One selling point was that its waters were fifty times more radioactive than the strongest known natural source, at Gastein in Austria. The prospectus blithely claimed:

> The German Company's radium preparations have the effect of rendering water radioactive to any degree desired, and the radioactive water can then be therapeutically applied in the form of drinking water or baths, or by inhalation; in the same manner as natural radioactive water of much lower strength is applied at the various Continental Spas. When so applied the water is of the greatest use in the treatment of diabetes, albuminuria, arterio-sclerosis, myocarditis, neuritis, neurasthenia, rheumatism, gout and spinal cord troubles, and in fact all diseases of which lowered vitality is a feature. [35]

Radium Ltd itself suffered terminally lowered vitality when it was forced to close down by order of the High Court in London in 1918 under the powers of the Trading with the Enemy Act. Enthusiastic advertising by companies such as Radium Ltd persuaded a susceptible public but, worse, encouraged the medical charlatans, who began to recommend radium injections for a range of complaints, both real and imaginary. One such 'specialist' promoted the same radium treatment to cure both schizophrenia and something rather colourfully called 'Debutante's Fatigue'. Even the much-publicised death from brain abscesses of a famous US playboy tennis star who had taken daily doses of radioactive water failed to frighten the horses. Radium hysteria continued, and George Bernard Shaw pronounced that:

> The world has run raving mad on the subject of radium, which has excited our credulity precisely as the apparitions at Lourdes excited the credulity of Roman Catholics. [36]

By the autumn of 1910, the new radium plant in Limehouse was about to produce its first radium bromide. The pitchblende and assorted tailings had begun to arrive from the Trenwith Mine in June, and by the end of August 500mg of crude bromides entered the fractionation process. Towards the end of October, the first half-gram of radium bromide was securely installed in its lead and asbestos safe. This was the occasion of a tour of the works by invited visitors and journalists. Ramsay told the gathering that he expected to get about 500mg from each ton of Trenwith ore, and that he was still confident of producing a gram each month:

> The supply of radium is thus assured. From a medical point of view alone the demand will be very great; in fact the present demand is greater than the supply. [37]

At the time the total quantity of radium in the world was thought to be about five grams. Ramsay said that further experiments aimed to obtain uranium, polonium and actinium – all of which were present in the Trenwith pitchblende, which Ramsay described as being much richer than that of St Joachimsthal.

In 1911, a second British radium producer appeared. John Stewart MacArthur was a metallurgical chemist from Glasgow whose discovery in the 1890s of the cyanide process had been responsible for saving the world's gold and silver extraction industry from economic decline. (His technique increased extraction rates from 40 to 95 per cent and is still in use today.) MacArthur set up his first radium plant in the village of Halton at Runcorn in Cheshire. Gas Street – as its name implies, the home of the village's gasworks – ran from the main thoroughfare, Bridge Street, to the bank of the Bridgwater Canal; the site had previously been used as a lead smelter. This was a pragmatic decision based largely on the fact that Runcorn was the seat of England's expanding chemical industry, which would be capable of supplying the needs of the chemically intensive extraction process. MacArthur, always concerned to find the means of harnessing technology for public benefit, was intrigued by the new science. He attracted a group of the young chemists who had already worked with him around the world, and encouraged them to become proficient in the new processes, intending that Britain should develop its own indigenous production.

John MacArthur had recently been involved in the establishment of the Porcupine Gold Mines, at Timmins, Ontario, about a hundred miles from the Québec border. This was a large development by Consolidated Gold Fields of South Africa. It was one of the deepest mines in Canada, and is still economically significant. MacArthur's interests in geology enabled him to develop an appreciation of the wide range of uranium minerals from different countries. He knew that pitchblende was available in Cornwall and Portugal, but he studied North American minerals in particular, and it was while at Timmins that he decided to enter the new industry. Autunite and pitchblende ores were available for exploitation in Gilpin County, Colorado, but the most prolific, carnotite, existed in substantial amounts in both Colorado and Utah, where it had been used by the Ute and Navajo Indians to provide red and yellow uranium salts for body-paint and the decoration of buckskins. MacArthur studied these minerals, and the possibility of a new direction in his own work:

> Carnotite is a complex ore consisting essentially of vanadium existing as oxide, or with potassium as double silicate and associated or loosely combined with uranium oxide. It occurs chiefly in the Rocky Mountains over a wide area extending about 100 miles on each side of the northern Colorado–Utah boundary. The ore occurs in pockets in a sandstone or shale country. The country

> abounds in fossils of recent date, and this has suggested that the carnotite is too 'young' to permit of the radium having attained its full equilibrium value. The mathematical computation for this period based on data still far from complete is about 10,000,000 years. In point of fact the uranium–radium ratio is little different in carnotite from other ores of unimpeachable antiquity.[38]

> The figures involved in the radium industry baffle the imagination. As an ordinary carnotite ore assaying about 2.5 per cent uranium-308 contains radium equal to only about 12.5mg per ton, we cannot expect to extract more than a maximum of 10mg per ton from such ore, which is one part per 100 million. The human mind cannot grasp this figure; it cannot even grasp one million. We all understand a day or an hour as a unit; one million days would take us back to the times of Nineveh and Babylon, one hundred million hours would take us back more than 11,000 years, and to extract 10mg from a ton is like selecting 10 minutes out of the Christian era. [39]

Radium production was going to present formidable problems and the prospect of becoming involved in formulating new processes and technologies was enormously attractive to him. The restless need to make the kind of innovative breakthroughs that he had experienced as a young chemist at the Tharsis Sulphur and Copper Co. in Glasgow had not deserted him. MacArthur purchased carnotite concessions in Utah and set about the practicalities of mining the ore and transporting it across the Atlantic. He became a director of the International Vanadium Co. of Liverpool, which had purchased Blackwell's Metallurgical Works Ltd at Garston in 1909, and began extracting vanadium for steel-making. Most of the directors were based in Baltimore, New York and Pennsylvania, and the company established the General Vanadium Co. and the Radium Extraction Co. as American subsidiaries. [40]

By 1914, several of the large London hospitals were trying to use radium, but all complained of the expense and problems of supply. The Cancer Hospital, the Middlesex, the Metropolitan, the London Hospital, and St John's Hospital for Diseases of the Skin all reported success in a letter to *The Times* but 'we are all wanting radium badly – very badly – but we cannot afford to buy it.' [41] Simultaneously, the public was getting hold of the wrong picture. *The London Daily Press* talked of 'radium for all' [42] – in much the same way that Britain was told about 'electricity too cheap to meter' at the time of the opening of the first nuclear power station at Calder Hall in 1956. In the over-excitement, it was inevitable that cynical voices were also raised against radium. The *Mining Journal* published an article under the heading 'Radium Dreams', which poured cold water on the whole subject:

> We see how far we are still from the passage out of radium dreams into the region of fact. We get the impression that the discovery of Mme Curie only tantalised humanity. [43]

The article went on to insist, not unreasonably, that success would only be achieved when governments put enormous scientific effort into finding uranium ores, and when Stock Exchange speculators, 'who have only obscured and discredited the question', were removed. But the radium business was to find it difficult to divorce itself from the sheer fun of the new substance. The owner of the biggest production company in the USA, a one-time undertaker and alleged hawker of patent medicines named Joseph Flannery, established a house journal in 1913 which promoted considerable work on the internal administration of radium. One of his employees, C. Everett Field (who had received all of eight weeks' training before being sent out to promote the use of radium), expressed concern at the prospect of being questioned by medical professors. He was told by Flannery: 'Answer them any way you choose – nobody knows enough to refute you.' [44]

Field took the advice and went on to make the recklessly bold assertion that:

> Radium for several years has been given internally and by injection in large doses with absolutely no disturbing symptoms. It is accepted as harmoniously by the blood stream as is sunlight by plant life. [45]

In 1926, Field stated that in the previous twelve years he had administered 6,000 intravenous radium treatments.[46]

The sunlight analogy was extremely popular, no doubt driven by the exciting luminous property of radium. In the early 1900s, two prominent US doctors, Robarts of St Louis and Morton of New York, were obsessed by the possibility of using radium to induce fluorescence as a therapeutic agent within the body, and added fluorescent substances to radium salts for that dubious purpose. With the supposed efficacy of their 'Sunshine Therapy' and the measurement of the strength of radium being widely advertised in homely 'Sunshine Units', they were well on their way to full-scale quackery. Robarts died of excessive radiation exposure in 1922 at a time when such a diagnosis, even if accurately made, would never have been allowed to see the light of day.[47] The battle between the spivs and mainstream medicine for supremacy in exploiting radium was not to be settled quickly; how could the public react with anything other than enthusiasm when respected medical journals published confident, informed opinion?

> Radioactivity prevents insanity, rouses noble emotions, retards old age, and creates a splendid youthful joyous life.[48]

A long series of skirmishes was to take place between opposing factions, while impartial researchers continued to try to understand the implications for mankind of this new and unbelievably powerful element.

FOUR

'AT THE DISPOSAL OF RICH AND POOR ALIKE'

As the First World War loomed, British academics and metallurgists were active in trying to secure supplies of radium ores and the means of processing them. At the same time, others were intent on building the appropriate organisation to further academic research and ensure the reliable supply of materials for medical use. In both cases, self-sufficiency was the goal. During 1913, John MacArthur was training his new ensemble of chemists by the side of the Bridgewater Canal in Runcorn. Despite some evidence for the beneficial effects of the medical uses of radium, which he hoped to facilitate, he was also conscious of contentious claims and had observed the growth of radium spas in particular with a sceptical eye:

> It was possible to obtain radium-water, mud, or vapour baths at various Continental spas, which established 'emanatoria'. The treatment was used mainly for rheumatic subjects and neuropaths. It is hardly to be wondered at that the multitude of adventitious aids to bodily and mental well-being which are to be found at such resorts tended to obscure the real nature of the curative powers of radium.[1]

Europe was in a state of great political instability, with an arms race already running out of control. At home, Ireland and the Suffragettes had a virtual monopoly on the domestic news. As John MacArthur experimented with the practicalities of trying to establish production on something like an industrial scale, the sources of ore supply were about to be closed. No country was able or willing to admit that it had anything other than meagre native mineral reserves, whose exclusive exploitation had to be protected. With Cornish reserves extremely limited, the necessity to import ores into Britain became a major issue. Sources in Utah and Colorado still remained open, although transport across the Atlantic in times of approaching war was fraught with doubt, danger and expense. The uncertainty in the situation could be brought to an end at any time by the total prohibition of the export of such materials. MacArthur had at least reached a conclusion as to how to proceed:

> The treatment of complex ores is necessarily complex, and radium ores are no exception to this rule, and there is no standard method of treatment. As radium is analogous to, and by all purely chemical tests indistinguishable from, barium, it suffices to treat an ore as if one wanted to extract barium, which, generally speaking, is converted to carbonate by treatment with carbonate of soda, the carbonates thus formed being dissolved in hydrochloric acid and separated from most of the soluble constituents by precipitation with sulphuric acid. This precipitate contains, besides the radium, the barium and lead contained in the ore. If barium is not contained in the ore a small portion of a barium compound must be added. The other constituents of the ore, such as uranium, vanadium, and bismuth, are dealt with by ordinary laboratory methods applied on the industrial scale.
>
> Finally, one has to deal with a mixture of barium and radium sulphates, the former in overwhelming excess, say one part of radium to 100,000 of barium. The mixture of sulphates is solubilised by

> carbonating as before, dissolved in hydrochloric acid and crystallised. It is found that when such a solution is saturated at the boiling point and allowed to cool, it deposits half of its barium, which contains four-fifths of the radium. This fractionation is repeated time after time, passing the crystals forward and the mother liquors back, so that each lot of back-going mother liquor meets the lower lot of forward-going crystals to form a new solution for a new crop. The net result of a long series of such fractionations is that we get at the upper end a small fraction of say 100mg of almost pure radium chloride, and at the other end a very large fraction of crystals of barren barium chloride.[2] *2 (*Mother Liquor:* the solution remaining after a stage of crystallisation)

In the spring of 1914, discussions began in Dublin on the prospect of forming a radium institute; in Manchester a Radium Fund; and for Scotland, the Glasgow & West of Scotland Radium Committee. These bodies (and others in Hull, Portsmouth and Sheffield) all aimed to model themselves on the success of the Radium Institute in London, which was spearheading both acquisition of regular supplies of radium, and its proper distribution to hospitals and universities. In Swansea, where a commercial company was set up with the same aims, one of the directors was Sir Alfred Mond, the industrialist and liberal politician son of Ludwig Mond.[3] In Manchester, the industrialist Sir Joseph Whitworth and the brewer Sir Edward Holt were foremost in campaigning for funds. Manchester took the Holt Radium Institute so much to its heart during the First World War that German Street was renamed Radium Street in its honour. The Holt Radium Institute, which was amalgamated with the Christie Hospital in 1932, is still very much alive (and Holt's brewery launched a new beer in 2003 with the promise that 20p from every pint sold would be donated to the Institute). It was the ability of organisations such as the Holt to raise funds that enabled infirmaries to equip themselves with radium; for, as its therapeutic value appeared to increase, so did its cost. However, they did not become involved in arranging the actual treatments – their sole functions were to raise funds and acquire the supplies of radium.

In Glasgow, the driving force behind the small group which decided to secure the future of radium for the city was Sir John Stirling-Maxwell of Pollok House. Stirling-Maxwell was one of the wealthiest men in Scotland, and an arboriculturist and art collector (his Pollok estate is now the setting for the prestigious Burrell Collection). He had a life-long friendship with the physiologist (Sir) Walter Morley Fletcher, who became in 1914 the first secretary of the Medical Research Committee (the forerunner of the Medical Research Council). Stirling-Maxwell's small group of supporters included Frederick Soddy, Sir Donald MacAlister, principal of Glasgow University, Sir George Beatson of the Glasgow Cancer Hospital, and representatives of the three Glasgow infirmaries. A public meeting attended by the city's most prominent citizens was held in the Merchant's House, which proved that there would be enough wealthy sponsors to kick-start their plans. Stirling-Maxwell began by stating a case against radium, and followed this with a full and imaginative scientific case in support. Finally, in promoting the case for pure philanthropy, he told the meeting:

> It is our boast here that all that science can do is at the disposal of rich and poor alike when they are ill. I wish to state the case moderately, but I confess it makes me shudder to think that in a disease which claims so many lives, the rich citizen who can afford to go elsewhere for treatment, or to pay to have it brought to him, now possesses a hope of recovery which remains quite out of reach of the inmates of our hospitals.[4]

Stirling-Maxwell's egalitarian ideal was well-received in Glasgow. His scientific exposition was also perfectly judged. But as he concluded his address with a hint of caution, even he could not resist a nod towards the new 'magic':

> The indiscriminate use of such an agent is not to be thought of. It would be dangerous to life and utterly useless for the advance of science. A complete record of its use and a reliable diagnosis of the cases to which it is applied are absolutely essential. Nothing but disappointment can be expected otherwise. The radium will be in use so far as possible night and day. Its energy is incessant. It needs no rest. But by a wonderful contrivance its energy can be stripped from it and conveyed to a distance. The receptacles loaded with this separated energy – the emanations as they are called – possess for a limited time the power of the radium itself. They seem to have come out of a fairy story or a book of magic – tiny tubes of glass not bigger than a match whose contents are invisible in daylight but glow with a white light in the dark, so fragile you can snap them with your finger, so powerful they can burn into your flesh.[5]

By the time the first formal executive committee of eighteen members met a month later, over £5,000 had already been donated to the campaign. One of the executive members, Dr John MacIntyre, of Glasgow Royal Infirmary, noted their decision:

> A physicist acquainted with the subject should be appointed to keep the radium, to make and advise in the making of different kinds of applicators, according to the surgeons' wants. Further, it is important to have medical men who are skilled in the use of radium to apply it to the patients, and each hospital could make its own arrangements for doing this.[6]

He went on to make interesting reference to the idea of continuing 'rental' of radon emanation, rather than account costs against the 'one-off' and expensive therapeutic use of pure radium bromide itself:

> The expense of working the radium might prove to be something between £300 and £500 a year but, as the emanation tubes could

> be hired out for private work and charges made for them, as is done in the Radium Institute in London, a certain revenue would be obtained each year.[7]

A sub-committee was set up to arrange the details of purchasing and controlling the use of the first 300mg. The first meeting was held on 13 May and there was an immediate bombshell – Frederick Soddy resigned from both committees. He was on the point of leaving Glasgow for a professorship in Aberdeen, but it seems that he had been displeased that not enough emphasis was being placed on the scientific investigation of radium. The committee boldly refused to accept his resignation:

> It was remitted to Sir George T. Beatson to approach Mr Soddy and assure him that the scientific aspect of the value of radium as well as its value in treating malignant disease has always been recognised and will be kept prominently in view by the committee, and to confer with him regarding the conditions under which he would be prepared to give the committee his service.[8]

Soddy stuck to his decision, but rather grudgingly agreed to act as advisor. He was asked to arrange the purchase of 300mg of radium, and this time the inducement of a fee of 100 guineas was sufficient to activate him. Soddy knew John MacArthur and was aware of his activities in Runcorn. He negotiated with him for the supply of hydrated radium bromide (not fully fractionated) at £10 per milligram. This was a very good price for the committee, but the 300mg would be in a mass of material weighing about 15kg. To be pure enough for medical use, further fractionation would be required in a Glasgow laboratory. To undertake the later stages of fractionation, Soddy recommended his assistant, Alexander Fleck (who later became Lord Fleck of Saltcoats, and the second chairman of ICI). In the light of MacArthur's generous price, the committee agreed to increase their purchase to 600mg. They also agreed to appoint Fleck as physical chemist, at an annual salary of £250.

In June, the committee decided on the terms of their proposed use of radium. In the first place, as was common at the period, it would only be used on wholly inoperable cases; it was seen not as an alternative to surgery, but as a treatment of last resort. Treatment was to be made available only through the five hospitals represented on the committee, and on the following terms:

1. Each institution making application for radium must be solely responsible for the treatment.
2. A radiologist should be appointed by the institution, who should direct the treatment in every case.
3. Clinical and, as far as possible, pathological evidence should be furnished regarding the nature of the cases proposed to be treated.
4. No operable cases of cancer should be subject to radium treatment. [9]

On 6 August 1914, the Glasgow Radium Committee took delivery of its first radium bromide from Runcorn, and MacArthur was paid a first instalment of £2,014 6*s*. For handling the material, a radiometric laboratory, with a lead and asbestos radium safe, was provided in the Chemistry Department at Glasgow University for use by Fleck during the final fractionations. The Radium Committee had good reason to be satisfied with its first 'deal', since the price of radium worldwide had been steadily climbing as its scarcity became clear and the range of potential uses increased. Another reason for a touch of self-satisfaction was that Glasgow had achieved its goal with the civic dignity of its most prominent citizens intact. The Manchester Radium Committee had raised three times the amount of cash donated in Glasgow, but only by orchestrating outlandish popular appeals in the tabloid press. The funds of both cities were small compared to the huge sums raised by the Radium Institute in London, which benefited from royal patronage and a rather 'glitzy' approach designed to attract money from the king's racing chums and other celebrities.

By 23 February 1915, the Glasgow Committee recorded the supply by MacArthur of the final delivery of radium bromide, and the Minute Book records that, 'the contract price of £6,000 had been paid and the contract fulfilled to Prof. Soddy's complete satisfaction.' At the same time, Alexander Fleck made a report of his fractionation of the radium bromide from Runcorn:

> The fitting of the new laboratory was completed in late September and work on the fractionation was commenced early in October. The 600 mgs. of radium bromide was received in eleven lots delivered at various times between 5th October and the following 8th January. The total weight of material containing the radium was 28.876 kilogrammes and the average content is therefore 20.8 milligrammes of radium bromide per kilo. I am therefore in a position to supply a considerable amount of radium emanation for hospital purposes and I would suggest that 300 mgs. be set up for the collection of its emanation and that I proceed as rapidly as possible to obtain the remaining quantity. [10]

About this time a second mine in Cornwall was re-opened with hopes of obtaining radium. The Union Mine (known as South Terras) was situated at Tolgarrick Mill, in the River Fal valley near St Austell. The mine originally provided iron and small amounts of tin, but the Uranium Mines Co., which exploited South Terras for ceramic-colouring materials for processing in Germany, failed in 1891 although the mine itself staggered on under different owners for some years. After the discovery of radium, pitchblende was identified at a deeper level in the mine, and the existing surface waste dumps were also deemed worthy of exploitation, thus avoiding having to deal with one of the biggest problems at the site – water penetration. The mine had never been pumped, and when the workings became flooded, work had to stop until water levels subsided. While some material was supplied to the new London Radium Institute, most of the ores were shipped to a radium processing plant that opened in 1912 at Gif-sur-Yvette, 18 miles from Paris.[11] In 1904 the *Cornish Guardian* published a series of articles on the potential significance of radium to South Terras, but relied on the use of one of the popular but meaningless illustrations:

> The three great salient features of radium are its enormous energy, heat and light. So prodigious is the velocity of the electrons which radium emits, that if the total energy of one gramme were converted into weight-raising power, it would be able to raise the whole of the British Navy to the top of Ben Nevis. [12]

South Terras was taken over in 1912 by the Société Industrielle du Radium Ltd (a British-registered company whose parent, Société française du radium, owned the plant at Gif) with the intention of increasing the commercial exploitation of radium. A total of £152,500 was paid for South Terras, including associated mines at Tolgarrick, Resugga and Carwalswick.[13] Mining operations at South Terras were suspended on the outbreak of the First World War in 1914, and it was to be some years before developments were restarted.

Early in 1915, MacArthur decided to abandon his radium operation in Runcorn. His main difficulty had been obtaining clean enough water supplies, and a return to his native Scotland would solve that problem. His in-laws owned land and other properties at Dalvait near the village of Balloch on the River Leven, about half a mile from its exit from Loch Lomond, and it was there that a vacant sawmilling yard was found by the river bank. MacArthur retained the lease of the property in Runcorn, which he continued to make use of for the refining of antimony. *The Times* reported his move:

SCOTTISH RADIUM FACTORY: AN INDUSTRY FOR LOCH LOMONDSIDE

Dalvait on Loch Lomondside is to be the home of a new radium industry in Scotland. An old sawmill has been adapted for use as a factory for the extraction of radium and other rare metals for their ores, and within the next few weeks a start will be made. The promoter of the scheme is Mr John S. MacArthur, a Scottish metallurgical chemist, whose attention was directed to the problem of radium extraction about two years ago, when he began to make industrial experiments, working with ores from which uranium and vanadium had already been extracted. He was soon successful and before long had established a small factory at Runcorn, where he ultimately employed about two dozen assistants, the majority of whom are now expert in laboratory experiments. The difficulties at the outset were very great. He had to do practically all the work with his own hands while training his assistants. Though Runcorn was perfectly suitable for most of the early processes through which the ore passes, for the finer and more delicate processes a purer air and water supply was necessary, and it was this which influenced him in his decision to transfer his factory to Loch Lomond.

Already Mr MacArthur has placed upon the market about 1,500 milligrammes of radium, part of which was purchased for use by the medical profession in the district, by the Glasgow and West of Scotland Radium Committee. In the new establishment at Dalvait he expects to be able to extract uranium and vanadium, while there will be other products, chief of which is a radium fertiliser.

The factory will be the first of its kind in Scotland. So far as is known the only similar enterprise in this country is being worked by a London concern which depends for its supplies upon Cornwall. [14]

The rectangular site was about 45m along its short western edge flanking the River Leven, and occupied about 1,400sq.m. The main processing shed measured about 27m by 11m; it was a large, single-storey building with a high-pitched and louvred roof on the south side of the site, with a small boiler-house at right-angles to the rear. On the north side of the site MacArthur's in-laws made a £500 contribution to the new company by building a smaller single-storey block with windows, occupying about 23m by 8m, to be used as offices and a laboratory. Away from the river, a cart track led up a slope to the nearby roadway.

The River Leven is fast-flowing, and the water at Dalvait was extremely clean, although further downstream it was heavily polluted by the huge textile dyeing and printing industry of the Vale of Leven. There was a nearby railway station, which provided good links with Glasgow for the delivery of chemicals. As he moved into the old sawmill, MacArthur gave an interview to *The Scotsman* newspaper:

> We hope to found an industry which has hitherto been in the hands of foreigners, and to make it a permanent thing. Of course it is not possible even to indicate what the demand for radium may be in the future. Its possibilities are so enormous that everything will depend on the development of the directions in which it may be applied. It will be a new industry for this country, though not novel in itself. The production of radium has formerly been confined to Austria, Germany and France. Austria had the ore, and worked it, but Germany and France had to buy it. Germany was far and away the biggest producer: we got all our supply from them. Before the war we sent all our crude production to Germany, who refined it, but now we propose to do everything ourselves.
>
> Radium certainly seems to be expensive till one comes to consider the matter closely. Fifty milligrammes cost £1,000 but that would be sufficient for the treatment of 100 persons a year, which means only £10 a head: but as the life of radium is many centuries, one need not consider anything beyond a rental, and assuming that the rental is 10% of the value, the treatment of 100 patients would cost £1 each. In calculating the price, one has to

> remember all the circumstances of the production of radium. The ore is not plentiful. A great part of it comes from the remote Rocky Mountains, and before it lands in this country it has actually cost about £20 a ton for transport. Then its extraction involves, say, 50 delicate operations, and though the ore contains one sixth part of a grain to the ton, it is not safe to reckon on getting more than one grain from ten tons. [15]

MacArthur hoped to be able to increase his output to about 5g or 6g a year. Besides radium products for medical uses, he was also interested in extracting uranium and vanadium for use in high-grade steel and armour plating. One other possibility which intrigued him was the production of fertilisers using varying amounts of radium ores and residues. This matter had been receiving a great deal of attention in Europe and America. The previous year, *The Times* had reported startling effects on the germination of seeds subjected to alpha particles generated by radium. Suttons & Sons, the seed merchants, conducted a long series of tests at their experimental grounds at Reading. Radishes and lettuces grown in radium fertiliser had been particularly successful and mustard seeds, for example, had shown a 63 per cent increase in germination rate. [16]

MacArthur brought most of his laboratory staff, including William and George Dempster, his chief chemists, with him from Runcorn. His niece Sheina Marshall, who later went on to become a noted marine biologist, also worked for a time in the laboratory at Balloch. Other laboratory staff, boilermen and office staff were recruited locally. In April 1915, J.S. MacArthur Ltd was formed to operate the Loch Lomond Radium Works. William Dempster was a director, as was James Nairn Marshall, one of MacArthur's brothers-in-law, and his nephew John Smallwood MacArthur was company secretary.[17] An office was opened in Glasgow, at 180 West Regent Street, an address shared by the Radium Products Co., which marketed their materials.[18]

By 1915, the situation in the USA had changed dramatically. Whereas there had been early interest in quite small-scale extraction of uranium for ceramics and vanadium for steel, there was less enthusiasm in American companies for the local extraction of radium. There had been a number of very small plants attempting to extract uranium and carnotite in Gilpin County, Colorado, but it was not until 1911–14 that commercial-scale chemical extraction began. The Bureau of Mines founded the National Radium Institute in Denver, with mining operations in Paradox Valley, Montrose County, Colorado, and laboratories in Denver. The Bureau of Mines regulated all mining, refining and export activities in America, as well as conducting experimental laboratory work on behalf of the US government. However, most of the new private companies were fiercely independent, and made it their business to try to poach chemists from St Joachimsthal and elsewhere in Austria and Germany. The new hysteria over radium recalls the famous 'railway mania' that occurred in Britain in the 1840s, and would be repeated years later when uranium became the 'must have' mineral resource.

The Standard Chemical Co. of Pittsburgh (established in 1911 by the entrepreneurial undertaker Joseph Flannery) was the first major company set up principally to produce radium, and it remained dominant until prices collapsed after the First World War. Carnotite was mined in Paradox Valley, Colorado; it was then carted sixty miles across country by rough stagecoach track to Placerville, for mechanical treatment. The crushed ores were transported to a converted stove factory in Canonsburg, Pennsylvania, where initial processing took place in ordinary domestic porcelain bathtubs. Partly-processed materials were then carried by hand on the public tramcar system to the company's Radium Research Laboratory in Pittsburgh, for final fractionation. The company claimed to be 'the largest producer of radium in the world', with three times the total production of the rest of the world combined. Although Flannery was regarded as a maverick, having previously sold dubious patent medicines, he entered the radium business after losing his wife to cancer, and had very likely responded sincerely to the distressing consequences of that dreadful illness. Standard Chemical established a subsidiary, the Radium Chemical Co., to market its products, which it did from premises in midtown Manhattan before moving in 1955 to larger premises in a single-storey building in Woodside, Queens, where a wide range of radium and other radioactive substances were prepared for medical and industrial use until as late as 1988.[19]

The Radium Co. of America, in Sellersville, Pennsylvania, produced its first radium salts in late 1913. On a much smaller scale than Standard Chemical, it was defunct within about five years. The Carnotite Reduction Co. mined ore in Colorado and extracted radium at a plant in Chicago; and the Chemical Products Co. had a refining laboratory in Denver. A number of humbler concerns were mining relatively small volumes of ores, largely for export, although the war in Europe was curtailing the markets both for ore and refined radium salts. There was concern in the privately controlled radium industry that the US government – which was conducting affairs in Denver in great secrecy – might try to impound or otherwise take control of all mining areas; or even worse, completely nationalise radium production in order to drive out speculators, as the research director of Standard Chemical feared:

> Government production of anything is notoriously slow and expensive, and it is my opinion that the worst thing that could happen to the radium industry and the science of radium therapy would be the governmental production of radium in America. The delusion of cheap radium would delay the acquisition of adequate supplies of radium by hospitals, and radium treatment would be sadly hampered, and by cutting off the market in the United States, producers would be forced to sell their radium abroad, or go out of business. [20]

In Scotland, John MacArthur first got underway at Dalvait with the production of radium salts for medical use. One elderly lady, who as a child took lunch boxes to her father who was employed at the Loch Lomond Radium Works preparing fires for the acid vats, remembered the sense of busy activity at the laboratory. Five men were employed in preparing the ore – drying, grinding and turning. There were four large boiling vats, each 30ft long and 5ft high. Six men were employed tending the vats. They were supplied with face-masks and wooden clogs, since the acids destroyed leather boots – even the steel tackets on the soles. The workers were apparently made aware of the possible dangers of the chemicals with which they were working. The vat house was fitted with roof-louvres for ventilation, and it is reported that escaping fumes were sufficiently toxic to kill birds which perched on the roof. Radium bromide was prepared for insertion in 'quills' – 'just like blotting-paper rolled up ... about the size of a match'. The quills were wrapped in cotton-wool and heavy-gauge lead, and secured in a small, lead-lined box which was then chained and padlocked to the arm of either MacArthur or William Dempster for delivery to Glasgow hospitals, where the quills were used in the preparation of radon emanation. My informant remembers her father stirring vats of boiling acid, and recalls that at one point 'he was advised to get out, because of the burns on his hands'. Similar advice was later apparently given to George Dempster, as 'the flesh was all coming off his hands'.[21]

Despite the difficulties and uncertainties of production (not to mention the unrecognised dangers of the new technology), the First World War was to make positive as well as negative impacts on the radium situation. Although the USA and other mineral-rich countries were effectively shutting down, military demand for luminous materials ensured that as soon as hostilities ended, commercial production of radium made new progress. Meanwhile, the timely establishment of the various radium institutes guaranteed that medical uses would be able to continue during wartime.

Medicine seemed like a 'safe bet' for radium, but even here, misinformation was a major factor. It was understood that radium seemed to behave like X-rays, and might therefore have similar properties. This was correct, up to a point; gamma radiation, because it has no mass or charge, is deeply penetrating, and in broad terms is similar to X-rays. However, radium also emits alpha and beta particles, both much less penetrating but with the potential to be very much more dangerous, especially if ingested. Early erroneous medical belief was that not only could radium (i.e. the gamma radiation) burn away tissue, but that (magically!) it could selectively destroy diseased cells and encourage the growth of healthy cells; this latter claim is still supported by adherents of 'hormesis', an effect not widely believed at the level of therapeutic doses. Another claimed benefit was that radium could destroy bacteria, making it immediately attractive as the wonder-cure for the scourge of tuberculosis.

The initial excitement over radon and radium-water was accompanied by references to 'the stimulation of cell life', to 'feelings of well-being and contentment', to 'complete lack of ill-effects' and the wonderful 'fairytale action'. While medicine would indeed make beneficial strides, the main effect of such purple prose was to draw in the charlatans; where medicine was cautious, the entrepreneurs were to be reckless. With their arrival would come the true growth of the radium industry, as efforts were made to incorporate radium into all manner of products for casual unsupervised use in the home. All the interpretations of radium were socially and (oddly enough) commercially positive; but the priesthood of science would have no monopoly.

FIVE

'THE BURNING BUSH OF MOSES'

The celebrated surgeon Sir Frederick Treves – famous for his humane treatment of Joseph Merrick, the 'Elephant Man' – was interested in becoming a practitioner in radium, but wanted a tight rein on early over-enthusiasm. In 1909 he gave a lecture at the London Hospital, in which he warned that not enough was known about which diseases were appropriate for treatment:

> It was essential to ascertain the action both of the rays and of the emanation on bacteria and their products. Then the selective action of radium on the tissues must be investigated. What would be the effect of introducing it into the substance of a growth in tubes made of material that was permeable to its rays? What were the effects, local and general, of radioactive water? In Paris such water was sold as an application for indolent wounds; and at Joachimsthal a large spa was being erected to enable the public to drink it. In conclusion the lecturer repeated his warning that extreme caution must be exercised in speaking of what radium was going to do.[1]

One London supplier of radium materials obviously thought the same, and decided in the autumn of 1912 to avoid the populist use of terms such as 'Sunshine Units' and to try to be more scientific in describing its products. The commonly used method had been to describe the strength of radium preparations in terms of millions of 'uranium units', but Hopkins & Williams of Hatton Garden decided that all their products (which typically contained 90–92 per cent radium bromide) would now be sold on the basis of the actual quantities of radium bromide that they contained, using a method certified by Marie Curie.[2]

At the beginning of January 1913, after its first full year of operation, the London Radium Institute issued its first report, in which it gave a summary of 657 cases treated by radium or radon. There were 38 cases examined but not treated; 41 had been so recently treated that results were not yet available; 39 received prophylactic irradiation only; 53 were apparently cured; 28 were cured; 245 were improved; 70 were not improved; 88 abandoned treatment, and 55 had died. It was pointed out that many of those treated were already extremely ill, having exhausted other resources of surgery and medicine, and would therefore be generally regarded as otherwise beyond improvement. The only cases refused treatment were those in which the patient was practically moribund, or where the disease was regarded as unsuitable for radium therapy. The Radium Institute by this time kept its stock of radium in 'an iron room resembling a huge safe'. Radium was being applied using over forty different pieces of equipment, designed specifically for the treatment of cancers in different parts of the body. These applicators were also designed and supplied for use in hospitals in London, the USA, Denmark and Germany.[3]

A further enthusiastic report came from the Institute in October that year. The medical superintendent, Mr A.E. Hayward Pinch, and Sir Frederick Treves announced that they had 4g of radium in their possession – thought to be more than half the total world supply. The Institute was now completely in favour of the use of radium 'emanation', and were using 1.5g of radium solely for the production of radon, which was supplied in sealed tubes. A series of hollow plates had also been designed, into which radon was forced.

> If a doctor in Edinburgh, for example, wants 200 milligrammes of radium for use upon a patient, its cost, £4,000, probably would be prohibitive; but the Institute can supply a plate containing radium emanation which will have the same effect for an amount which is comparatively trifling. Radium gives off the emanations constantly and itself is not destroyed. It is the only reproduction of the burning bush of Moses – constantly giving off heat and never consumed. The activity of the emanations, however, when fixed in a hollow plate or tube, decreases, falling to one-half strength in three and a half days. [4]

In the previous ten days, radium emanation equivalent to 860mg of radium costing over £17,000 had been supplied to other institutions, and radon equivalent to 150mg of radium was being despatched to hospitals daily:

> So far as we have ascertained, this is the only institution in any country which has produced a hollow emanation-containing plate and has distributed emanation in this way. [5]

Rather incredibly, the Institute was also supplying radioactive water for internal consumption by patients, as Treves enthusiastically reported:

> The improvements brought about in the condition of patients by drinking radium-water strong enough to be luminous are marvellous. Usually a patient drinks about a half-pint of the emanation water daily for six days in each of six weeks. That is the first course, and after a rest the course is repeated if necessary.[6]

At least this risky procedure was being carried out under medical supervision. Charlatans were also at work, and their activities attracted the attention of the authorities, as *The Times* reported in April 1914:

> In an article in *The Times* of Wednesday attention was drawn to the fact that substances supposed to possess radioactivity – but which in fact are quite inert – are being freely purchased by a too-credulous public. [7]

As well as sanctions against fraud, the paper called for regulation within the British Pharmacopoeia and the Food and Drugs Act. It also warned against indiscriminate use of materials which actually did contain radium:

> That the public should swallow radioactive water, even if fully attested, without medical direction, is undoubtedly a matter of grave concern. Those who feel any inclination to play with this dangerous material should understand clearly that they do so at their peril. [8]

As events in the USA would later prove, when the public was able to indulge in the casual, unsupervised use of radium-water, the results could be disastrous; matters became even worse when the public was enabled to infuse all manner of materials with radioactivity. Treves observed in his report that the Institute was staying open for sixteen hours a day in order to satisfy demands on its services. However, it had to close completely for a month each year to give some rest and recovery time to staff, 'all of whom have upon their hands burns caused by radium'.

The following month, Sir William Ramsay spoke about the production and supply of radium for hospitals at a meeting of the British Radium Corporation. He wanted to increase the production of radium and extend its use to a wider field of medicine than had been attempted so far.

> A great many of the hospitals had acquired small quantities of that substance, but so far only a few diseases had been attacked in that way. In certain cases success had been attained, but there were a great many diseases which had not been attacked, as nobody had had the time, and there had not been a sufficient supply of radium.[9]

Ramsay enthused about their extraction method, which he said was being constantly improved. He also drew attention to the fact that their process gave rise to the prospect of the medical use of other substances that were released:

> There were other substances in the Trenwith ore and, indeed, in all radium ores, which had not so far been exploited from the therapeutic point of view. They were polonium, ionium and actinium. They were at present chemical curiosities which had not been extracted in any great quantity, though there was no very great difficulty in extracting them, because their works at Limehouse had so far been fully occupied in dealing with the radium contents of their ore.[10]

He pointed out that, in his view, radium itself could not be used on humans because of its expense, its danger, and because of the fact that it was not readily eliminated from the body after administration. However, he felt that polonium could achieve a similar beneficial effect and be eliminated after about three months. Like Treves, he was also keen on the commercial prospects for radium-water:

> There was a very big business in selling radium emanation, or niton, dissolved in water, to private patients, medical men and hospitals, at very fair prices. Indeed these emanations represented, in a concentrated form, those which occurred in most of the spa waters at places such as Bath, Buxton and Carlsbad.[11]

Despite the efforts of the Radium Institute to manage supplies of radium for all the main London hospitals, the governors of the Middlesex Hospital wanted to reach an agreement with the Council of King Edward's Hospital Fund for the more economical provision of radium. They were convinced of the therapeutic value of radium in treating localised cancer, but were aware that results were often still contradictory, and continuing research had to be conducted, as Sir Alfred Pearce Gould pointed out in late 1913:

> They had those results because they were ignorant of its power. With increased knowledge of its power, what were now to them inexplicable freaks regarding its use would become wholly simplified matters. For that reason he trusted that, whilst they used radium in the clinical department of the hospital, a certain small proportion would always be allocated for scientific work in the cancer laboratories for extremely important and necessary researches into those emanations. [12]

One unexpected manoeuvre adopted by some institutions with regard to safeguarding radium supplies was insurance protection. Some hospitals were insuring fractions of a milligram per year, despite the high costs and the fact that not even the insurance industry was sure quite how such a procedure would work in practice. One complicating issue was that it was alleged that radium was being misappropriated by patients with, quite apart from the loss of valuable assets, potentially very dangerous effects on whoever interfered with the material. *The Times* reported the opinion of 'a distinguished man of science who is an authority on radioactive substances' as saying that such insurance could only be seen as a very hazardous business, as quantities were so minute, they could be easily mislaid and were virtually incapable of identification:

> The quality of any particular piece of radium could be ascertained only by a long process of experiments. It was usually kept in the form of varnished plates or as emanations. Its insurance was quite a different thing from that of gold or diamonds, which were identifiable or at least capable of being valued easily and expeditiously. The risk of adulteration and the difficulty of immediately discovering it, made the insurance of radium a 'desperate business'.[13]

In America, the US Bureau of Mines was becoming concerned that, while little radium was refined there, the country supplied two-thirds of the world's uranium ore. It was noted that the deposits of ore in Colorado and Utah 'are being rapidly depleted by the enterprise of foreign scientists'. Several individuals set out to promote an indigenous American industry. The Colorado gold magnate Thomas F. Walsh was one. As early as 1908 he had exhorted graduates of the Colorado School of Mines:

> We are only at the beginning of knowledge of the deposits and the powers and properties of the rarer minerals. Radium, that miracle of the century, will illustrate what I mean. We know but little of its properties, but even that fills us with wonder and amazement, and hints at undreamed of possibilities in this unexplored region of physical science.[14]

The US National Radium Institute was set up in 1913 by the Bureau of Mines, James Douglas, a New York industrialist and philanthropist, and Howard Kelly, a deeply religious gynaecologist from Baltimore, who was a faculty member of Johns Hopkins University Medical School and who ran his own private hospital. The Institute was established to obtain radium from native carnotite ore and to supply hospitals in Baltimore and New York. A large radium refining plant was established at Denver, Colorado, and a 'radium bank' for US medical uses was to be set up at Johns Hopkins University in Baltimore. In parallel with these moves towards consolidating the infant industry, there developed a strong, but controversial, bureaucratic elite which favoured the approach of the Austrian government – state control and the banning the export of uranium ores completely. [15]

In December 1913 reports came from Baltimore that surgical success was being achieved; a thirty-year-old US congressman from New Jersey (and personal friend of President Woodrow Wilson) named Robert Bremner, who had been born at Keiss in Caithness, at the most north-easterly point of the Scottish mainland, where his family were fishermen, had been successfully treated for cancer at Kelly's private hospital:

> Tubes coated with rubber and containing radium to the value of $100,000 [£20,000] were inserted in incisions made in the extensive cancerous growth in the left shoulder. Yesterday the second application took place, and the tubes were allowed to remain in the growth for twelve hours. [16]

Despite the massive doses used on his celebrity patient, Kelly complained of a shortage of radium (to which he normally attributed biblical powers), and he persuaded the *New York Times* to give repeated positive publicity to the Bremner case ('Bremner's progress gratifying' 3 January 1914; 'Bremner sends message to public' 12 January 1914; 'Plea for Radium cure' 7 February 1914). Unfortunately, Bremner died in February 1914, and even the usually enthusiastic Joseph Flannery claimed that his death had been caused by excessive amounts of radium; in addition, the Maryland authorities debated whether Kelly had violated ethical medical procedures by his aggressive promotional activities.[17] Despite these potential setbacks, the issue had generated public attention, and Kelly felt that the publicity had justified his support for the proposal made at the end of December 1913 by the US Secretary of the Interior stating that the federal government should appropriate all radium-bearing land. Scientists were apparently outraged by this proposal, but doctors and hospitals (and certainly Kelly) supported such a move on the grounds that supplies of radium for therapeutic purposes should not be at the mercy of 'land-grabbing trusts and speculators'. [18]

By early 1914, advances in the chemical processing of radium were made in Britain by Kent Smith and H.B. Rolfe, and in the USA by Col. R.B. Moore, a chemist working for the Bureau of Mines in Denver.[19] In Austria, work was underway at what was then the village of Neulengbach, about 20 miles from Vienna. This quiet backwater, characterised at the time as a place full of retired officers and snooping neighbours, had its moment of fame when the Austrian expressionist Egon Schiele was arrested there, accused of posing children for erotic drawings. There, Prof. R. Sommer and his colleague F. Ulzer perfected and patented an improved method of treating carnotite, a mineral that they had previously regarded as generally uneconomic.[20] Their colleague Prof. E. Ebler also devised a final extraction process which avoided the longer fractional crystallisations.[21] Another result of the shortage of rich uranium and radium minerals was that various attempts were made to develop the use of mesothorium (^{228}Ra), which had similar properties to radium (^{226}Ra) but which had a half-life of five and a half years (compared to that of radium of 1,600 years) and therefore required a more rapid replenishment of stocks. Mesothorium was generally available as a by-product of the gas-mantle industry, being produced during the processing of imported monazite sand.[22]

The combative Dr Walter Lazarus-Barlow, director of the cancer laboratories at the Middlesex Hospital, was a highly outspoken contributor to the radium debate. He was convinced from his own work that radium was of undoubted value in treating cancerous tumours, but was reluctant to overstate the case until he had studied the outcomes over a longer period of time. He did not want to contribute to the climate that fostered constant talk of new cures being announced in quick succession on the basis of insufficient understanding. He was a bitter critic of the established circumstances of radium production and supply, and said it was shameful that radium could not be more readily obtained, saying: 'I have every reason to believe, from the statements of manufacturers, that radium could be sold at a profit at a few shillings a milligramme. The present market price is £20 and upwards.' He was strongly of the view that the state should take control of the supply of radium on behalf of the 35,000 people who died from cancer annually in Britain: 'this ridiculous inflation of price can only be artificially engineered, and it is little short of a disgrace that the economic law of supply and demand should be strained to the detriment of suffering humanity'. He alleged in early 1914 that the price of radium was inflated and that the producers were operating a 'corner' to keep the price high. The producers denied any such deliberate distortion of the market, and legitimately claimed that not only were the appropriate grades of ore hard to find and acquire, but also complicated and time-consuming to process; in addition, some countries, initially Austria but now probably also America, were banning the export of the ores precisely because they were so scarce. Lazarus-Barlow was unmoved:

> Patients are turned away daily, and the radium the hospital has is booked up for a month or six weeks ahead. There are 92 beds constantly filled with inoperable cases. It is heartbreaking. [23]

The chairman of the British Radium Corporation, the civil engineer Sir Francis Fox, issued a statement saying that the corporation wanted to enable the hospitals to have secure supplies of radium. However, while they had been holding substantial stocks with that in end view (with little response from the hospitals), German buyers had stepped in to buy the entire stock at higher prices than the hospitals would have paid. In order to meet the ever-increasing demands from British hospitals, the hint was dropped that the corporation intended to build additional new works.[24] Norman Whitehouse, the Radium Corporation's chief chemist, claimed that Lazarus-Barlow's comments were misleading. He pointed out that at the time the corporation had substantial stocks of almost-pure radium bromide,

> I made a personal tour of most of the hospitals in London, offering the same at £14 12s 6d per milligramme. That the Middlesex Hospital did not avail themselves of this opportunity of securing all they wanted was entirely their own fault or that of their adviser. A month or two later large quantities were sold by us to important German purchasers (including the city of Frankfurt), both for immediate and forward delivery, at enhanced prices. [25]

Surprisingly, the official method of detecting radium ores in the field at this period harked back to the sometimes accidental actions of Henri Becquerel twenty years earlier. The guidance issued to prospectors by the Denver office of the US Bureau of Mines said:

> The best way to test these ores is to wrap, in the dark, a photographic plate in two thicknesses of black paper. On the paper lay a key and then, just above the key, suspend two or three ounces of the ore, and place the whole in a light-tight box. Pressure of the ore on the key and plate should be avoided. After three or four days, develop the plate in the ordinary way; and if the ore is appreciably radioactive, an image of the key will be found on the plate. [26]

The intentions of the House Committee of Mines to restrict the exploitation of radium ores and their processing were hardening, given the prospect of the war in Europe and the likely increased foreign demands on American resources. In 1914 the only two companies in the country manufacturing radium were the

Standard Chemical Co. of Pittsburgh, and the Radium Co. of America in Sellersville, Pennsylvania. The former had a lucrative contract for supply of radium to Europe, which made heavy demands on public land identified as containing pitchblende ore.

> How far legislation could prevent this process is problematical, as a good deal of radium land is already in private hands. There is also opposition on the part of Colorado, and other Western states, which dislike Federal interference. One is inclined to suspect that the real trouble is not so much the want of regulation as the fact that Europe has so far been ready to outbid the US for radium and radium ores.[27]

The claims of the government that there was likely to be a radium 'famine' in the USA were vigorously rejected by Standard Chemical, whose director of research insisted that the USA was producing far more radium than the whole of the rest of the world. He estimated that his own company conservatively expected to produce up to 24,200mg of pure radium bromide during 1914, and half that amount again the following year; their output alone for 1915 would equal half the world stock of radium.[28] Speaking as the self-interested capitalist that he was, he insisted that with government production of *anything* being notoriously slow and expensive, the delusion of cheap radium would hamper the supply for hospitals and become a luxury for the few.

> The war abroad will cause no radium famine in America – and if the radium industry is to grow – if radium is to be produced from American ore in America for Americans as the Secretary of the Interior and the Directory of the Bureau of Mines have been urging – more can be accomplished now by encouraging private capital to continue in this field than by discouraging it by such legislation as is now contemplated.[29]

His Standard Chemical colleague Warren F. Bleecker was even more forthright in extolling the virtues of their business and in damning any form of government control (which he characterised as being specifically designed to hand over the radium industry to favoured individuals):

> During the last two years, certain persons have prevailed upon a branch of the national government to assist them in entering the radium business, not in competition with, but to the ultimate exclusion of, the producers now in the field. So far as the general public know, this is purely a government enterprise. My understanding is that it is not purely a government enterprise, but an arrangement on the part of private individuals, for government backing, in order to obtain a monopoly of the radium business.[30]

His thesis seemed to be that only Colorado could supply the entire world, and that if Colorado's resources were 'wasted by crude and improper methods of treatment' then the entire usefulness of radium would be lost for a generation. He did not believe the government prospectus that by taking control of both the mineral resources and the subsequent processing the price of radium would fall significantly, thereby making radium more available for therapeutic purposes.

> To those who may have been misled by more or less recent statements, that the ores now mined in Colorado are for the most part exported and thereby lost to our home people, it may be said that the small amount of ore which was mined by prospectors, up to the time the Standard Chemical Co. began operation, was sent to Europe. Since that time, as a matter of record, at least 85% of the carnotite ore mined in south-western Colorado has found its way to Pennsylvania for treatment. It would only be justice to the capitalists and engineers who have put their energies into

> the solving of this problem to state that the Standard Chemical Co. is now producing four-fifths of the world's supply of radium. No mention has been made, I believe, in the Bureau of Mines' publications as to the magnitude, either actual or relative, of this strictly American enterprise.[31]

Bleecker's complaint against 'certain persons' appears to have been aimed at the National Radium Institute. However, when the Institute's plant in Denver was closed in 1916 it was confirmed that its stock of radium had been produced for about a third of the cost that Standard Chemical were charging.[32] As events unfolded, neither side won the bitter battle. Ore deposits were not nationalised, and in any event turned out to be more extensive than anyone thought. Prices of radium did not fall for about twenty years, by which time most mining and processing activities were taking place on the other side of the world.

While the Standard Chemical Co. was claiming that the radium industry was in a vigorous state, others pointed out that at the end of 1913 in the entire USA there were perhaps only three or four surgeons with any real experience of using radium; there was also a prescient warning that the subject should remain firmly within the field of mainstream medicine:

> In the progress of the future applications of radium to the curing of disease, nothing is more to be feared than its use in nostrums of every kind. The 'wonders of radium' have been so extensively exploited in the public press that already the name is being employed as a physiological agent in advertisements of all kinds of materials, many of which contain no radium at all or, if this element is indeed present, in such small quantities that no therapeutic value can be expected.[33]

What certainly happened as war engulfed Europe was that, regardless of the lobbying of commercial interests, the production of radium in the USA rocketed, as did its use in medicine and in a bewildering range of products. Both the Standard Chemical Co. and the Radio-Active Chemical Co. of Denver, among others, were experimenting with radium fertilisers for horticulture; this seems likely to have been an appealing activity largely as a means of marketing otherwise useless residues with unattractive disposal costs. Initially, the gardens and lawns of those who owned and promoted the industry were the beneficiaries, and there were reports of increased growth in trees and vegetables. Soon, as with the activities of Suttons at Reading in England, more elaborate testing was undertaken, using mainly corn and soya beans, at experimental farms in New Jersey, Illinois and Ohio, supervised by academics from the Universities of Columbia and Illinois.

In Glasgow, the Radium Products Co. was curiously listed in the trade directories under 'manure manufacturers'; or perhaps not so curious in view of John MacArthur's long-standing interest in radium fertilisers. His main product, 'Aurora Radium Fertilizer', was sold in five quantities from 7lb (at 1*s* 6*d*) to 1cwt (12*s* 6*d*); lots of 2cwt and over were despatched free by rail from Glasgow or Balloch. Research on both sides of the Atlantic was undecided on the subject, but work continued with high hopes of success in some areas of horticulture. After two years, results in Illinois were frustratingly uncertain, with one result for radium for each one against:

> In all of these trials the average variation from the checks is so slight and so evenly distributed, 'for' and 'against,' as to lead only to the conclusion that radium applied at a cost of $1, $10 or $100 per acre has produced no effect upon the crop yields either the first or second season. Radium, with all its wonderful energy, is found upon careful analysis of the known facts, to afford no foundation for reasonable expectation of increased crop yields, when financial possibilities are considered. The rate of application would cost about $58,800 per acre at present prices for radium.[34]

The Illinois researchers calculated that the heat evolved by a thousand dollars' worth of radium on an acre of land in a hundred days would be less than the heat received from the sun on one square foot in thirty seconds!

Research was also taking place in Rome, at the New York Botanical Gardens, and the Royal Society in Dublin. Ireland's capital had been an early centre of experiments, with work being carried out using radium bromide on cress seedlings at Trinity College as early as 1903.[35] In December 1909, the journal *American Naturalist* had reported experiments into the use of radium in stimulating the germination of lupins and timothy grass; these had produced the confusing conclusion that extended exposure produced an initial beneficial effect followed by total inhibition of growth. Everywhere, the results were confusing. However, since trials involved the annual germination and growth of a wide range of seeds and plants, considerable time was required for assessment; such factors as genetic change caused by radiation were not considered in such short-term trials. There was evidence that the use of radium residues was more successful than radium salts or emanation. Likewise, the improved germination of seed appeared to be a more successful target than that of the enhanced growth of established plants, or any other effect such as herbicide action. Some bulbs seem to have been amenable to radium fertilisers; the Denver Radium Service had success with narcissus using radium-water as a stimulant, and John MacArthur showed comparative 'with' and 'without' photographs of hyacinths which appeared to indicate considerable achievement, as described in *Scottish Country Life*:

> Hundreds of tests which have been made on various plants show that, when radium is applied in small quantities, the germination is accelerated in the proportion of two to one against normal soil; that at the mid-season of growth the radiumised plant maintains its lead, though it does not increase it, and that at fructification

> the radiumised plants weigh up to fifty per cent more than the plants grown under normal conditions.[36]

One of the great benefits of using radium fertiliser was apparently that 'it never needs to be applied again'. In September 1915, Martin Sutton of Suttons Seeds invited scientists and journalists to inspect the results of the Reading experiments. They had worked with rape, radishes, tomatoes, potatoes, onions, carrots and marrows, and the conclusions were similar to those of Illinois University:

> While in some cases plants dressed with radioactive ore had given better results than the control plants, the improvement had not been of such a nature as to warrant the assumption that so expensive a commodity as radium could be profitably applied to crops. [37]

At a meeting of the Royal Society in London, the industrial chemist Thomas Thorne Baker gave a lecture, during which he discussed at length the use of radium-based fertilisers. He pointed out that there were many different mineral ores in use, and an even greater range of chemical processing techniques; indeed, he had been greatly confused in examining the circumstances of the Sutton experiments, which had used eight different residues, in addition to pure radium bromide:

> I found, on investigation, that the residues came from different sources, some copper and uranium, while one of the residues I happened to know contained no radium at all. [38]

Martin Sutton demonstrated specimens which he had grown using Baker's 'Nirama' brand, produced at his Carrickfergus plant in Ireland. Sutton confirmed that while overall results were difficult to quantify, germination seemed to be the most likely area for future success.[39] Baker insisted that attention was given to the influence of associated impurities in the ores, and to the much more complicated factor of the relative activity of the different rays produced by radium. He was convinced that the experiments so far showed two results; that the alpha rays were largely responsible for increased germination; and that emanation was of value to plant growth, thus suggesting that if radioactive matter was spread upon the soil, rain or watering would produce a beneficial effect by leaching. These results of 1915, although seemingly uncertain, did not stop research continuing. MacArthur maintained his own interest, and his fertiliser – a by-product of other activities – continued to be produced:

> Radium fertiliser must not be thought of as a food; it will not take the place of essentials such as phosphate or potassium, but will aid in their assimilation. It also seems to assist in the fixation of nitrogen. [40]

Evaluation of tests on radium-based fertilisers seem to have been scientifically rigorous, often more so than for materials intended for use on humans,[41] and despite the indeterminate results, the practical work continued throughout the entire period of the use of radium, and academic work long after that. In the 1930s unprocessed carnotite ore was sold as a fertiliser to citrus growers in Florida and California, and radium/uranium-derived soil supplements were still marketed in the 1950s.[42] The whole question is one which remains controversial to this day, and experimental work continues with modern radioisotopes into the effects of different forms of radiation on plant growth and seed germination. Much of this work is on a very small scale, but in 1987, 9 million hectares of farmland in China were producing improved crop varieties of rice, wheat, cotton, soya-bean and maize derived from the use of radiation-induced mutations, of which there were more than 800 available.[43]

This first attempted use of radium in an industrial field proved to be both confusing and disappointing, although others were to follow, with varying outcomes, before the all-embracing benefits of luminous paints were exploited. Medicine remained the principal focus of the radium industry, although even here the characteristics of the pantomime were to make more than a few undesired appearances. The melodrama would become worse; when the hysteria over luminous paint and the rash of pointless new products both increased, so did the number of physicians who decided to feather their nests by selling their scarce radium supplies to paint manufacturers and the succession of quacks who set up in business.

SIX

MEDICAL SLAPSTICK

The principal therapeutic uses of radium quickly settled into either of two forms. *External* application was used to irradiate tissue by the use of carefully measured and positioned radium sources contained within specially designed instruments – usually glass or metal tubes or 'needles' (or sometimes fabric soaked in radium-based varnish) taped or otherwise held in position for the required period of time. As knowledge was acquired through experience, the problem of blocking beta radiation while permitting the more penetrating gamma radiation was also addressed. Continuing improvements were also made in adapting methods of application to different parts of the body. *Internal* application involved controlled inhalation, ingestion, injection or similar means of introducing direct contact between radium source and tissue. Both methods involved the use of either radium salts or radon 'emanation' and a very great deal of experimentation, trial, error and the individual invention of new procedures and equipment took place. In most cases, the use of radon came to be the preferred method, given its relative cheapness and the fact that its 'half-life' was only 3.8 days (this meant that the exhausted radon source could be left *in situ*– subject to other clinical considerations- thus avoiding additional surgical trauma to a seriously ill patient).

Later, there developed an entire industry devoted to impregnating other items with radium. This began with the bubbling of radon through tapwater to produce 'radiumwater', which was widely sold for a range of medical and quasi-medical conditions. Devices were soon on the market to enable the gullible public to impregnate their food, drink and clothing with the health-giving magic of radium in the comfort of their own home, without the bother of having to defer to interfering doctors. Unfortunately, while the use of radon was inherently safer than radium salts, it was fairly rapidly eliminated from the body, so some doctors encouraged the almost continual consumption of radium-water in order to try to maintain 'an elevated level'.[1] That kind of over-provision was seen as a gift by the charlatans who were always willing to go further and faster than more considered medical opinion and practice allowed.There was yet another less medically approved field of activity which became a booming business – the promotion of radium springs to health fanatics. The 'strongest' springs in Europe were reported to be the Butt Springs in Baden-Baden, at Bad Kreuznach, at Gastein in Austria and the Italian springs at Lacco Ameno on the island of Ischia. In Switzerland, radium springs were developed at Andeer, Baden near Zurich, Fideris, Pfaffers, St Maritz, Solis near Thresis, Lavey les Bains and Discutis in the Rhine valley.[2] The later obsession in America was to undergo breathing of the radon atmosphere in abandoned uranium mines, and both there and in Europe, radium springs and radon mines continue to attract the credulous wealthy. But in the earlier days, radium 'balneology' was undoubtedly popular. Over half a million people took spa treatment annually in 1912 and the most famous of several establishments in the USA where radium-waters were favoured was at Springdale in Boulder County (where the locality is still known as Curie Spring). Others were Saratoga Springs in New York State, Mount Clemens in Michigan (where one clinician was candid enough to note that, '...much of the humbuggery of the medical profession still lingers about mineral springs. It is indeed a humiliating acknowledgement')[3] and the still-flourishing national park at Hot Springs, Arkansas. Hot Springs was the location of a major military hospital and when the federal government discovered in 1905 that the waters were radioactive, it was assisted in their promotion by the US Railroad Administration:

> The waters [at Hot Springs] are radioactive, and by means of the bath every rheumatic joint, every sealed-up pore of the skin may be not only reached and cleansed of impurities, but renewed under the influence of that brain-baffling curative which we call radioactivity... More than ninety per cent of those who have taken a full course of baths have been either cured or benefited by them.[4]

In 1912 the 'radium industry' was well underway in Britain. Sir William Ramsay, widely in demand as a consultant, was called in along with staff from University College Medical School, and the analytical laboratory of *The Lancet*, to investigate the therapeutic value of the spa waters at the King's Well, the Cross Bath and the Hetling Bath at Bath in Somerset. Ramsay had been astonished and puzzled to discover that the Bath waters also contained 188 times as much neon as atmospheric air. Also in their sights were the Crescent Pump Room and the Gentlemen's Natural Baths at Buxton in Derbyshire, where the Hydropathic was already busy selling its radium-water ('strengthened' by the imported potion from Germany; the concept of tinkering with naturally occurring radium-waters in order to turn their fleeting radioactivity into a permanent benefit was receiving much attention). Although one professor in New Zealand was claiming that radon was 100,000 times more active than radium, most opinion suggested that radon was 'none too strong' and would benefit from being concentrated or 'strengthened'. In Ramsay's experiments, mice were fed with bread and cheese 'spiked' with radon, and kittens were injected, to determine how long the gas remained active.[5] There was also the question of how efficiently the skin absorbed the radon. After the initial investigations of Thomson, Dewar and Strutt a few years earlier, the King's Well at Bath was thought to be ripe for development and the proposition was made that the use of electricity would assist the absorption of radon by the skin. *The Times* reported the doubtful procedure:

> If therefore the patient in the bath were connected with the negative pole of a battery at a potential of 100 volts or even more, and the other electrode were placed in the water... a considerable dose might be given.[6]

This bizarre recommendation for the mixing of water, electricity and radium was well-received by The Hot Mineral Bath Committee and the City Council. The Radium Development Syndicate (director, Sir William Ramsay) obtained permission to go ahead with their project, at a cost of a quarter of a million pounds (about £10 million at today's values). The Pump Room's brochure described Ramsay's 'Radium Inhalatorium', where mineral sprays cost 2*s* 6*d* and radon inhalation, 3*s*:

> The radioactive gas which acts as a tonic and germicide may be inhaled through the nose or by the mouth, or it may be used with the mineral water in the form of nasal and throat sprays. Eye sprays have also proved efficacious, especially in gouty iritis. Aural douches are also given, special care being exercised. Striking results have been obtained in this department in cases of anaemia, chlorosis, and in all cases of rheumatic laryngitis and pharyngitis and in toxaemias, such as rheumatoid arthritis.[7]

A powered spray system already in use was thought to be suitable for giving a dose twenty times that of the open bath – even more if the patient was again connected to a large battery! The Inhalatorium was part of the larger Bath Hot Springs treatment area where, to the music of Reg Fowles and his Dance Band, the impressionable could sample a total of over sixty different treatments. They included such exotic species as Faradism, Schott Movements, Galvanism, and the Bourbon-Lancy Bath, with prices ranging from sixpence for the Scottish Douche to twenty-five shillings for the Bergonie Treatment. Mr T. Pagan Lowe, one of the medical staff at the Royal Mineral Water Hospital at Bath, who rejoiced in the feeling, '... that we balneologists are like the man in Molière's play who was delighted to find that he had been talking prose all his life', wrote with enthusiasm about the new development:

> The discovery of radium with its far-reaching effects on every part of the human economy has altered our views fundamentally with respect to balneological treatment. A more scientific era has dawned. Time was when a suspicion of a man's professional integrity was raised if he called himself a balneologist; but now, instead of groping in the shades of night, our landscape is illuminated by the sunshine of science and we march along with firmer footsteps and with head erect. Many of us believe that in this element we have found the explanation of the therapeutical effects of mineral water, but beyond a declaration of our faith we dare not at present go. Is it, then, any wonder that we balneologists regard radioactivity as the unknown god?[8]

That 'scientific sunshine' again. Within weeks, as the scientifically curious and the entrepreneurially opportunist scoured the land, radon was discovered in the ancient hot springs at Clifton. However, some clinicians were willing to recognise that 'contra-indications' showed up in the course of their activities. T.P. Lowe at Bath was concerned that colleagues working at radium springs and in radium refining in Germany were collating both subjective and objective symptoms. He was one of those who advocated caution:

> We should not reject it [the action of radium] merely because we are unable to explain it. For we must remember that there is still a great deal about radioactivity and its therapeutic action that we are unable to explain.[9]

Other fields for the practical use of radium were being developed by Thomas Thorne Baker, who had already investigated radioactive fertilisers. Baker was interested in the potential for destroying bacteria in food, and was commissioned by the Scottish Fisheries Board to produce radioactive salt for preserving fish. He undertook these experiments at the works of The International Salt Co. at Carrickfergus in County Antrim. The company was soon renamed The Radium Salt Co. and after developing its industrial salt, the principal domestic product was 'Ra-Ba-Sa' radium bath salts:[10]

> ... several officers who have come back from trench life at the front suffering from acute rheumatism have been treated with highly satisfactory results. It is hoped that this salt will, to a large extent, do away with the necessity for visiting Continental bath resorts... [11]

In addition to the destructive effects on bacteria, Baker was convinced that smaller amounts of radium could also have the opposite effect – of stimulating bacterial growth and fermentation, in much the same way that milk turns sour during an electrical storm:

> Tubes of milk stood in highly concentrated radioactive matter, the ß and γ rays penetrating the glass, but in each case there was very little result. If, on the other hand, milk is slowly filtered through a layer of insoluble radioactive matter, and the milk bacteria are

> brought into contact with the radiation, a very rapid effect is noticed, the milk clotting within a short time. In other words, radium is able to take the place of the ferment rennet, and may therefore prove of value in the manufacture of cheese.[12]

Baker was also attracted by the potential for developing new methods of applying radium for other antibacterial and antiseptic purposes such as mouthwashes and toothpastes; and in luminous materials, photographic emulsions and radium-water.[13]

The British Radium Corporation also celebrated a successful year, with plans announced for an even bigger radium factory having five times the capacity of the Limehouse Works. The new plant was under construction across the Thames at Croydon Road, Elmer's End, sited between a brick-works and the premises of the London Steam Carpet Beating Co.. But there was still genuine uncertainty among medical men such as Treves, and one of the explanations for what they saw as the ill-informed public confusion about radium was rather contemptuously referred to in the science journal *Nature* a few days after Treves' comments were published:

> It was assumed by the literary young men who write the leaders and notes in the daily papers that radium emanation had just been discovered, so they let their enthusiasm overstep the bounds of their knowledge.[14]

Another worry for the scientific community was that the quacks were beginning to gather. In November, *The Times* announced that the British Oxypathor Co., trading as The Radioactive Oxygen Institute of 62 Oxford Street, London, had acquired 3g of radium bromide from Paris- 'one third of the world's supply'. This company now changed its name to Radium Treatments Ltd, as it was expanding its ability to treat all manner of ailments: 'cases of Diabetes, Mellitus, Bright's Disease, High Blood

Pressure and Arterio Sclerosis have been found to do wonders under Radium Emanation'. What they failed to say was that they had been forced into the name-change after court action by Frederick Treves and the directors of the Radium Institute, who objected to the company's use of the word 'institute' in its advertising. There was obviously some justified ill-feeling that partisan commercial interests were encroaching on an area of what ought to remain specialised science. Under the headline 'Perpetual Energy' the company advertised boldly, with just a hint of mysticism:

> Some seventy-six years ago curative results obtained at Gastein and other spas were attributed to a gas contained in the waters and to nothing else, since they contained no mineral salts. There was then no method of testing the gas analytically; it appeared to be inert, or at any rate did not enter into combination with any compound or element. Experiments showed that it possessed the power of augmenting metabolism, increasing ferment action - peptic, lactic, pancreatic, diastatic and glycolytic. It also increased the excretion of urea and the elimination of uric acid, and produced markedly beneficial results. They called it WASSERGEIST, or the Spirit of the Water.[15]

Not only quacks were interested. Patent and company registrations of the period reveal that radium was of interest to purveyors of 'educational and scholastic articles', or 'scientific, optical and kindergarten toys', or 'articles of novelty, art and amusement of all kinds.' Sometimes, the word 'radium' was included in company names simply for its potent, somewhat mystical *cachet*. Some academics, too – however much frowned upon by their peers – were often tempted into the 'fun' opportunities of demonstrating the apparently magical properties of the new substance:

> An era when radium might be used as a fuel for ships and motor cars was hinted at by Mr C.E.S. Phillips at The Cancer Hospital, London. In solution, radium decomposes water, giving off an explosive mixture of gas. Mr Phillips produced three gases, and by fusing them with an electric spark, created a current of air which drove a small fan. Another experiment was described as a 'sporting event.' From the stage along the roof ran a glass tube 25ft long, connected with a tube of zinc sulphide. The lights were turned down and in the darkness radium emanations were pumped through the tubing and in an instant the sulphide was glowing luminously, while the audience – many of whom were medical men – loudly applauded.[16]

Phillips did a number of other party tricks with radium, and said that with coal stocks available for only 200 years, science would have to find a new energy source. Pointing at a speck of radium in a tube, he said, 'This may have enough energy to run a steamship across the Atlantic'. Newspapers in the USA were a great source of the most fanciful headlines about radium. One claimed that radium could be used to blow up battleships, while another announced the 'Secret of Sex found in Radium'. It was not only the more irresponsible newspapermen who waxed lyrical. Science fiction writers – even respectable ones such as Jack London – became obsessed with 'radium death rays'. In December 1911, an angry and frustrated reader wrote to the Journal of the American Medical Association:

> ... the newspapers are giving space to an account received by cablegram from Paris of what they call 'Radium Teas' – that is, exposure to radium emanations for a number of hours, during which time the patients play cards or read and take tea. Is this pure guff or is it an actual scientific process?[17]

The answer from the journal's Editor was that 'this purports to be a scientific procedure'.

The suspicion that America was about to embargo radium ores was well founded. The House Committee of Mines in Washington was still inquiring into the efficacy of radium as a cure for cancer and also into closing all land containing uranium ores. Legislation was not expected to be easy, however, since some states were expected to object vociferously on the well-worn grounds that they resented federal interference. Meanwhile, in Canada the Provincial Legislature in Toronto authorised the reward of $5,000 for the first discovery of radium ores in Ontario.

Doctors and surgeons had several rather bizarre problems in using radium. In one incident at Liverpool Royal Infirmary, a radium tube which had been taped to a patient's face overnight was found to be missing in the morning. Fearing that he might have unconsciously ingested it, doctors X-rayed him without finding the tube. There then followed a frantic tracing of the floor-sweepings from the ward. As this undignified process was being carried out, it was noticed that the municipal dust-cart was just leaving the infirmary. The Professor of Physics at Liverpool University was called in, and together with the hospital radiologist, he spent two days sorting through bins of municipal and hospital refuse till they eventually found the missing £1,000 radium tube.[18]

The high cost of radium encouraged unusual interest in preserving the tubes at almost any cost. *The Lancet* reported a case in Vienna in which a patient had accidentally swallowed a tube and was successfully operated upon for its removal. Not all such cases ended happily, though. In 1914, an unfortunate Preston woman unconsciously ingested a radium tube through a nostril. X-rays traced it, 24ft along her intestine. Emetics and other methods of removal having failed, she was operated on, but died a few days later. She had been given a complex anaesthetic cocktail of atropine, morphia and hyoscine, and an inquest found that she had in fact been poisoned. Inevitably, there were suggestions that she had been operated upon not for her own benefit, but simply to recover the expensive radium. However, the inquest was told that without the operation she would inevitably have suffered severe injury and probably death. The verdict of the inquest was 'death by misadventure'.[19] Practitioners do not seem to have improved their abilities to keep track of radium as time went on; in 1927 several thousand pounds worth of radium disappeared after an operating theatre at Charing Cross Hospital in London had been scrubbed. The building was searched, police were called and a private detective employed. He successfully discovered that the radium had been incinerated along with clinical waste and the detritus had been carted to a disposal site several miles away at Harrow. When the site was inspected with an electroscope, the radium was recovered, 'none the worse for its adventure.'[20]

The loss of expensive radon applicators often gave rise to events of low physics and high pantomime, as searches were conducted in likely places from trouser turn-ups to city sewers. One wintertime incident in Detroit resulted in a senior radiologist being carried slowly through the city refuse dump installed with his electroscope inside a recycled piano crate fitted out with a chair and oil-fired heater.[21] Another case of a radium needle having been accidentally thrown out with the hospital rubbish occurred in Souix Falls, Minnesota. Two physicists from the university were sent to the farmland dump 40 miles away to begin an interminable search. They experienced conflicting instrument readings, until they realised that a positive effect occurred only when a large herd of feeding pigs were in the vicinity. They had to undertake the misery of a close examination of 500 pigs before a process of elimination enabled them to summon a butcher...[22] Such incidents continued as long as radium was in use. As late as the last day of 1953, *The Times* reported that three radium phials had been erroneously flushed away at the Royal South Hants Hospital in Southampton, and that civil defence staff with Geiger counters ('used in the detection of radioactivity after atom bombing') were examining the public sewers from the hospital to the outfall.

A series of articles began to appear which made two things clear to the public; firstly, that medical science was not yet sure how the apparent efficacy of radium worked; and secondly, that its indiscriminate and untutored use was likely to be dangerous. There was increasing evidence that charlatans were entering the radium business. Headlines in the *London Times* told the story: 'Radium and Quacks: Medical Warning to the Public', and 'Radium Perils: Warning Against Quack Remedies'.[23] Medical extremists were abusing the existing, fragile state of knowledge; and more worrying, some were impostors who knew nothing at all. It was stressed that even medical experts did not know which of the radium rays- alpha, beta or gamma- had an efficacious effect and which might be damaging. With some medical zealots injecting patients with radium for complaints that they appeared to invent for the purpose, warnings were widely disregarded. It would be many years before the world became convinced.

Melodramatic stories – whether true or fanciful – were probably inevitable given the public's perception of the 'magic' of radiation. With the outbreak of the First World War – the first truly technological war, with whole populations menaced for the first time by the newfangled horrors of aeroplanes and poison gas – mad scientists and 'death rays' were becoming more real by the day. For those who did not themselves have the necessary imagination, H.G. Wells was on hand to prime the pump. But the real danger to the public came from the galaxy of products during the 1920s and '30s that contained radium – many of which remained on sale well into the 1960s. Charlatans, noting that doctors recommended the efficacious use of small quantities of radium, peddled the line that even greater quantities of radium would do even more good. The real swindlers, of course, made inflated claims about their radium products without actually bothering to use any radium. Like fraudsters the world over, they were interested in a fast buck and an even faster exit. In the USA one quack, James M. Harris of

Tulsa, Oklahoma, went around the country peddling what he called his 'Radium Oil' and extracting increasingly large sums of money from cancer sufferers. There is no evidence that his preparation contained radium in any form whatsoever. When the American Medical Association reported a series of court cases attempting to halt him, it concluded by lamenting that, '... meanwhile this bird of prey still continues his vulture-like activities.'[24] Other individuals went from state to state giving illustrated slide lectures, promoted by colourful publicity. Luther Gable ('lone survivor of a group who refined the first radium produced in America') was perhaps foremost:

> *The Astounding Story of Radium* is a graphic account of the life of the radium prospector, the mystery of radium laboratories, and the tragic death of his associates. Dr Gable also carries with him magic boxes, containing real radium. These magic boxes are perpetual motion fireworks machines and are passed out among those in the audience.[25]

From the time of the First World War there was a hysterical popular interest in radium, but not yet a full enough understanding of its dangers; there was also irrational political suspicion. In the USA, where radium production was increasing rapidly, there was paranoia that, in the aftermath of the war, European producers would form a combine to control world supply. But the US government's attempts to nationalise the industry were thwarted by the private companies, for reasons of profit and self-preservation; and by the individual states for reasons of antipathy to federal interference. It was widely asserted that government intervention would retard private philanthropy. Even as the First World War began, one multi-millionaire partner of the Scots-born philanthropist Andrew Carnegie offered to fund twenty specialist hospitals in America, each to be supplied with 5g of radium, at a total cost of $15 million. Given that the then total world supply of radium was estimated to be 30g, his promise was somewhat fanciful. His offer was never honoured, and was thought simply to have been a ploy in the opposition to the government's proposals.[26]

In Britain, where the industry was much smaller, there were similar uncertainties about the post-war situation. John Stewart MacArthur wrote about the effects on medicine in particular:

> A natural consequence of the war was that the progress of radio-therapeutics was much delayed through lack of workers and shortage of material. Less, therefore, has been done in the way of defining the possibilities and limitations of radium as a curative agent than might reasonably have been expected five years ago. In war surgery, however, radium has proved useful for preventing adhesions after wounds, and for rendering the scars pliable; in the case of facial wounds this has the effect of greatly lessening the disfigurement which they so often involve.[27]

With radium being introduced in new areas such as industrial testing and the preparation of luminous materials, there was a growing suspicion that medicine was losing the fight to secure supplies. Far from the American fear of European monopoly, there was in fact a complete lack of decisiveness about the future direction of the industry. MacArthur wanted sources of ores to be released and developed:

> Whether or not the production of radium is a key industry is a question which the mere scientist may think it presumptuous to answer; but it certainly ought to receive the attention of those who are devoting themselves to the solution of the problems of reconstruction. The position with regard to ore may be summed up as follows: the Joachimsthal pitchblende supplies, which showed signs of failing before the war, are now pretty well exhausted, and the United States government has placed an embargo on the

> export of radioactive ores. The British manufacturer of radium must therefore seek his raw material elsewhere. Cornish supplies are extensive, but require redevelopment. Promising reports of radium minerals come from India, South Australia, and East Africa. It is of immediate importance that those supplies be developed to their fullest capacity.[28]

In London in 1920, *The Times* voiced the worry about the place of medicine but suggested that the problem was being over-stated:

> The success which has attended the applications of radium to medicine, and the widely extended use of radioactive zinc sulphide for luminising watch dials, speedometers, compasses, and so on, has given rise to an annual demand for radium bromide which is greatly in excess of what was only recently regarded as the world's entire output. A few years ago a gramme of radium was looked upon as a prodigious amount; last year thirty grammes of radium bromide were produced by one company alone in Pittsburg. The fear has only recently been expressed that such large quantities of radium are being employed in the preparation of luminous compounds that a shortage of pure radium salts for medical work would result, but the quantities available and being actually manufactured are ample for both purposes, and a radium shortage is a remote possibility.[29]

It seemed the world had an insatiable demand for radium, but the people who were most sceptical – outside the area of medical science – were the producers themselves, who were unable to match supply to demand. The cost of production was still extremely high. In 1921, figures were published which cast an intriguing light on the relative values of radium and other international commodities. To obtain an approximate current equivalent of these values, multiply by thirty:

COMMODITY PRICE PER TON IN £ STERLING

Coal	1.5
Coal, as Diamond	400,000,000
Wheat	22
Cotton	124
Tobacco	1,700
Silver	6,000
Gold	208,000
Platinum	795,000
Radium	17,000,000,000 [30]

There were questions of ethics, too. In the post-war climate of national self-examination, there began a debate about The Role of The Scientist – a debate in which radium was a central feature. In a sense, the controversy began a few years earlier with the publication in 1913 of H.G. Wells' *The World Set Free*. Wells dedicated the novel to Frederick Soddy's important book, *The Interpretation of Radium*. Wells described the first splitting of the atom, a 'chain reaction' and, prescient as ever, actually used the term 'atomic bomb' (dropped from aircraft) for the first time. His story imagined a global war in 1956 in which 'puffs of luminous, radioactive vapour' almost destroyed civilisation. In the end, the world was saved and learned to live in harmony and universal love. This was verging on the mystical of course, but it also raised the moral stakes. It is said that in the mid-1930s the books of both Soddy and Wells inspired Leo Szilard to consider precisely how a chain reaction could become the force at the centre of the atom bomb. Soddy himself responded to the First World War with great anger, perhaps recognising what dangers the contribution of scientists (and in particular physicists, whom he rather despised) might generate. He became the first world-renowned scientist to attack the prospect of atomic warfare. Apart from his obsessions about political economy, he developed interests in women's suffrage, the Irish Question and taxation. In the end he was rejected equally by the scientists, whom he abandoned, and the economists, whom he tried to influence.

While the extreme dangers of radiation were not yet apparent, some of the early radium workers casually reported skin burns and scarring, and large volumes of radon must have been inhaled, generally without any immediate effect being recognised. Marie Curie became anaemic after many years of pioneering work and eventually died in 1934 as a result of her activities, but she had reached the respectable age of sixty-seven years. Another pioneer, Friedreich Giesel, who was the first man to produce radium on a commercial scale, did die of lung cancer, but not before reaching the decent age of seventy-one; and the head of radiological services at La Riboisière Hospital in Paris ominously died in 1933, 'following a series of operations and amputations'.[31] So, although radiation was having a deadly effect on many of the early scientists, that effect was often taking many years to become noticeable – so many years that the cause and effect correlation was not recognised; neither were potential genetic effects. Even the occasional publication of particularly graphic accounts of the effects of radiation damage made little impact. One early British radiologist, John Hall-Edwards, suffered severely from radiation injuries, noting at the time, '...the pain experienced cannot be expressed in words'. He later had both hands amputated.[32] In the USA, George Stover, an early pioneering radiologist in Denver who made numerous experiments with radium on his own body, was later found to have suffered grossly excessive radiation exposure which resulted in his having several amputations and over 160 skin-graft operations during the six years of his work with radium. Stover was a much-respected individual who properly analysed and recorded his activities in detail; displaying remarkable stoicism and restraint, he is said to have commented two years before his early death in 1915 at the age of forty-four, 'A few dead or crippled scientists do not weigh much against a useful fact.'[33]

There was well-developed concern centred around the rise in the USA of fraudulent misrepresentation and pseudo-science on the periphery of medicine. Some products placed on the market by untrained and uncaring 'businessmen' contained significant amounts of radium; others were accompanied by substantial claims without containing any radium whatsoever; in different ways, both were dangerous. Medicine was to have great difficulty extricating itself from the problems of supply, cost and the parasitical attentions of the charlatans. Too many 'advances' were being aggressively promoted by people who were little more than spivs, intent on parting gullible faddists from their money. As a later chapter will reveal, it was the enthusiasm that surrounded the invention and use of luminous materials that was to result in the first signs that such materials were to have a fearsome effect; even so, it would be some years yet before a clear understanding of the real nature of radium would emerge.

SEVEN

YANKEE ENTHUSIASM

In Britain, most radium producers still concentrated on supplying medical users, although John MacArthur had continuing negotiations with the Admiralty on the subject of supplying radium salts for the manufacture of luminous materials – principally paints for compass cards, gunsights and the like. Like wars everywhere, the outbreak of hostilities in Europe was responsible for many new industrial innovations, and the production of luminous dials was one of the biggest. The possibility had been raised of the Admiralty taking over the entire production of the Loch Lomond Radium Works, with John MacArthur himself retained as a consultant in charge. These discussions continued over several years, but MacArthur died before they reached any conclusion.

As far back as the 1870s there were already a number of commercially produced luminous paints, including Lennord's, and Mourel's, both using strontium carbonate; and Vanono's, using strontium thiosulphate. Surprisingly, these were not the earliest; a Professor Tuson in London had some 'of Canton's own make in a sealed tube, inscribed 1764', and there are accounts of other European manufacturers producing luminous paints in 1750.[1] Many of the technical magazines published in the 1870s in the USA had regular articles on luminous paint, and gave many formulae and instructions for their use in producing colour effects on different surfaces. One report in 1878 about a German product that could be used to coat wallpaper, which would then adjust its luminous effect according to the degree of light in the room, adopted a distinctly wry tone:

> What the country needs is not so much a paper that will grow darker in a light room, as one that will give light enough in a dark room to supersede the use of the costly and fickle gas and the fragrant kerosene. A contemporary remarks that any action, Yankee or German, that will light our houses by their wallpaper, will be received with becoming respect.[2]

Probably the most famous paint was that patented in 1877 by William Henry Balmain of Liverpool.[3] This, like the others mentioned, was not what the discovery of radium enabled – 'self-luminous' paint – but a phosphorescent compound using calcium sulphide (obtained by Balmain by roasting oyster shells to a white heat) mixed generally with a varnish containing salts of various metals such as zinc and magnesium. These paints required exposure to strong light to produce a limited and brief phosphorescence (although Balmain's patent application stated that full sunshine was not necessary and that daylight or artificial light need not be especially strong to produce an effect lasting for twenty hours). Balmain envisaged its use on watch, clock and instrument faces, signposts, signals, etc. More imaginatively, he suggested its use on buildings (allowing cost savings on the necessity of introducing artificial lighting) and in places, such as powder mills, where it would be dangerous to use any form of lighting which emitted heat or had other fire hazards. Most touchingly, he saw it as:

> applicable to the illumination of railway carriages by painting with phosphorescent paint a portion of the interior, thus obviating the

> necessity for the expense and inconvenience of the use of lamps in passing through tunnels. [4]

Balmain's paints were extremely popular, being used by manufacturers of other commodities as well as by the ordinary consumer at home. Luminous postcards were one popular souvenir, marked 'Treated with Balmain's Luminous Paint, and will shine all night!'; various recipes were also available for producing the ghostly photographic effect at home.

What was needed was 'self-luminous' paint, and radium was about to provide it. Many forms of dry radium salts naturally emit a soft blue luminous glow. Early radium workers would often take home a small amount in a glass tube, to demonstrate the effect to family and friends. However, the amount of radium required to maintain a constant visible luminous effect was so large – and expensive – that commercial exploitation seemed impracticable. After it was discovered that various substances became fluorescent when irradiated, experiments were tried in which small amounts of radium salts were mixed with various other materials. George Frederick Kunz, a young 'gemologist' with the New York jewellers, Tiffany & Co., found himself working with Charles Baskerville, a chemistry professor at the University of North Carolina. Baskerville seems to have been an averagely competent chemist who had stumbled into two specialised areas that were emerging as parallel interests to contemporary science – rare earths and radioactivity. Baskerville teamed up with Kunz to conduct wide-ranging tests on the extensive Tiffany & Co. collections of minerals and gems, classifying their phosphorescent properties under the effect of alpha, beta and gamma radiation. They perfected and received a patent for what was the first 'self-luminous' paint in 1903 by mixing radium carbonate, zinc sulphide and linseed oil, although it seems that neither man persevered with the commercial development of their invention.[5] Baskerville was one of many in the early days of radioactivity who either misunderstood or refused to accept the idea of radioactive transformation, and during his later work with thorium believed that he had discovered two new elements, which he named 'carolinium' and berzelium'. He became a falsely lauded celebrity in 1904 when the *New York Times* proclaimed him in a headline as: 'The Only American Who Ever Found a New Element'.[6] A similar patent was awarded in the USA to Hugo Lieber in May 1904, and the Ansonia Clock Co. of New York produced luminous products, almost certainly using radium acquired from Europe. [7]

A young American electrical engineer has also been claimed as the first person to produce radium-luminous paint. William J. Hammer was an assistant to Thomas Edison at his famous Menlo Park laboratories, where he specialised in lighting technology. In 1902 he visited Paris, and was a guest of the Curies, when they presented him with samples of radium and polonium to take back to the USA. He recorded the comment of a friend who had been seated between Bacquerel and Kelvin at a dinner, when the latter had remarked that:

> The discovery of Becquerel radiations had placed the first question mark against the principle of conservation of energy which had been placed against it since that principle was enunciated. [8]

Hammer was fascinated by that conundrum, and made its discussion a central feature of his many public lectures:

> Radium maintains its own temperature at 1.5° Centigrade above its surroundings, this being equivalent to stating that half a pound of radium salt would evolve in one hour sufficient heat to equal that caused by the burning of one third of a cubic foot of hydrogen gas; and that the heat evolved from pure radium salt is sufficient to melt more than its own weight of ice every hour. This evolution of heat, it is claimed, is going on constantly for indefinite periods and leaving the radium at the end of months of activity as

> potent as it was at the beginning. The problem therefore confronts the world of solving how radium can constantly throw off heat without combustion or without chemical change, as Prof. Curie says it does.[9]

Hammer acquired zinc sulphide and experimented with combinations which he eventually mixed with varnish to produce luminous paint for experimental application to a wide range of products, devices and equipment. He apparently did not patent his invention owing to the high cost of radium, but later became involved in legal disputes with other paint manufacturers in order to try to maintain what he regarded as his prior status in the field. He wrote and lectured on radium, and was interested in promoting medical uses for radium and radium-water. He also conducted research on X-rays, phosphorescence, fluorescence and 'cold lighting'.[10]

The onset of the First World War – not only the first 'technological' war, but one that would be fought as much at night as during daylight hours – was a real catalyst, generating a heavy demand for luminous paint for instruments of all sorts: compass dials, altimeters, watches, gun-sights, route marking tape and all manner of instruments that had to be readable at night without the use of excessive lighting, or the threat of mechanical or electrical breakdown. By that time, many different methods of production had been tried, and it was concluded that zinc sulphide was the most effective material with which to combine radium bromide. The zinc required very delicate preparation in a powerful electric furnace in order to produce crystals hard enough to withstand the intensely ionising and damaging bombardment of alpha particles from the radium, and information about this part of the operation was zealously guarded by manufacturers.

> The commercial material now available is guaranteed for the life of the instrument on which it is used. It is supplied to the

> manufacturer as a yellow powder. It is mixed with an adhesive and applied to the work at hand with the tip of a camel's hair brush. Care is taken not to brush the material but to place it on the surface drop by drop, so as not to crush the zinc crystals. It is interesting to note that the deterioration of the substance is not due to failure of the radium, but to a breakdown of the zinc crystals, due to the bombardment of radium particles. No known substance will stand up to this bombardment. [11]

The most long-lasting luminous effect was obtained when large amounts of radium were used relative to the zinc, but in such cases the zinc would quite quickly and completely disintegrate. The alternative involved using less radium and obtaining lower luminosity. The British Admiralty and the National Physical Laboratory at Teddington devised widely accepted international standards which recognised two radium values, the higher one (0.4mg of radium bromide per gram of zinc sulphide) for ship and aeroplane instruments and the lower for individual, personal instruments such as watches and compasses.[12] There was no universal acceptance of the new luminous paint technology, however. During 1915, British government scientists had sought alternatives to the use of zinc sulphide, which for unknown reasons appears to have been regarded as unsatisfactory. The conclusion of one Advisory Committee was quite grudging:

> Radium paint of the degree of brightness attainable is only suitable for tracing on instrument dials luminous markings of simple form, perhaps without any figures, and certainly without any small subdivisions. It cannot therefore be considered satisfactory as a general expedient for rendering instruments readable at night, but will probably be valuable as enabling roughly approximate readings to be taken when the ordinary system of lighting has broken down. [13]

In the USA, the zeal for luminous materials at all levels of society grew rapidly. The Standard Chemical Co. of Pittsburgh formed a subsidiary, the Radium Co. of Colorado, based in Denver; another Denver company, Schlesinger Radium, also formed a subsidiary, the Cold Light Manufacturing Co., to manufacture luminous paint for use on door plates, street signs, push-buttons and watch and clock dials; and the Radium Luminous Corporation was established in New York.[14] The application of the paint to watch dials and instrument faces seemed entirely routine and unexceptional; time was to prove it anything but benign.

Wartime restrictions had curtailed American ability to engage the European radium market, but the new fervour for luminous materials more than made up for that – albeit at the expense of medicine. The demand for radium for luminous paint became a crusade, with one estimate suggesting that 95 per cent of American radium production went towards luminous paint manufacture. In that climate, some American doctors contributed their own stocks of 'medical' radium for the purposes of paint production, in exchange for the appropriately inflated fee.

Clocks and watches were the items most commonly luminised. The US Army provided luminous watches for all its troops. In 1913, there were 8,500 in use, and six years later, the number was almost 2.5 million. Once America entered the war in 1917, the pace of production increased spectacularly. The Ingersoll Watch Co. alone acquired enough radium in Colorado and Utah to produce luminous watches at the rate of a million a year.[15] The various luminising companies vied with each other for the most vivid names for their products: 'Undark' was the fanciful name for the luminous paint of one company (of which, more later); others included 'Luma' and 'Marvelite'. Striking names were also promoted by the marketing departments of the prestigious watch-making trade:

> In the 'Radiolite' Robert H. Ingersoll has made a watch that does its work as well in the dark as in the light. It has hands and figures

> that glow like little stars in the darkness. They are coated with 'Radiolite' – a substance whose light-giving property comes from genuine radium.
>
> 'Radiolite' is independent of outside light; it gets its illumination from an element *within* itself. That element is radium. Every particle of radium, like a little battery, sends out energy in every direction. This radium – in infinitesimal quantities – is imprisoned within crystals of zinc sulphide, a phosphorescent material. The action of the radium on the crystals is to cause them to vibrate and glow, much as the filament of an electric light glows from the action of electricity.[16]

Advertisements by the Burlington Watch Co. ('practically every officer in the U.S. Army wears one.') featured a highly retouched photograph of a soldier consulting his Radium Dial luminous wristwatch in the cinema. Ingersoll also manufactured a range of 'Ingersollite' Locaters – luminous buttons and markers which could be fixed to keyholes, steps, doors, and any number of places only limited by lack of imagination – 'the little luminous tack that you can put almost any place'. Alternatively, you could buy 'The Little Spark you see in The Dark' – the Ingersollite Pendant, to be attached to light switches: 'Buy them today – and tonight you'll appreciate the handiness – the utility – of these little luminous guides!' The real fan could of course do it himself. An enterprising company in Chicago was only one of many who sold 'the very latest discovery in the scientific world', jars of luminous paint – in three sizes, from 25c to $1 – with the encouragement, 'Anyone – you – can do it.... Make your own Luminous Crucifixes, Rosaries, etc.' The promise was that 'THE DARKER THE NIGHT, THE MORE BRILLIANT IT SHINES!'. Even the military came up with as many silly devices as there were sensible ones. A popular American science magazine gave one absurd example:

> Our boys picked out this simple but effective invention to make startled Fritz jump: slabs of wood coated with radium paint. In the dead of night the slabs were stuck in the ground near German trenches, the luminous radium surface facing the Allies. Soon a nice fat snooping German silhouetted himself in front of a slab. In less time than it takes to tell, an American sniper had picked him off. [17]

The use of luminous paint on the humble military watch led within a couple of years to its use on all sorts of other items, from toys and trinkets to the human body. After the civilian automobile and aircraft industries adopted and modified the military uses of luminous instruments, the domestic knick-knack market pushed its way to the front of the queue. Religious objects, jewellery, dials and gauges, seat numbers, call-buttons, telephone mouthpieces, house number-plates, dolls-eyes, poison-bottle labels, fishing lures, costume accessories and make-up were only some of the items that received the treatment. Kits were sold enabling people to luminise their own clocks:

> Small flat discs treated with a radium compound are now being glued on the dials of clocks at the 5-minute points and also to the hands so that the clock can be read in the dark. A complete set of discs and a pair of hands can be affixed to a clock in a few minutes. The glow is practically everlasting and the discs, according to the manufacturer, will outlast the mechanism of the clock itself. [18]

Luminous paint was only the beginning; the radium industry was soon pulled into the realms of entertainment – and sometimes a world in which ideas that we now see as idiotic were probably promoted in all seriousness. Witness the chicken farmer who had the inspiration of adding radium to his chicken-feed hoping that it might provide self-incubating or even pre-cooked eggs.[19] It became the fashion for the most chic of New York hostesses to arrange radium demonstrations for the amusement of party-goers. Sir William Crookes had invented his spinthariscope to enable researchers to observe and quantify the scintillations produced by the disintegration of radium; needless to say, it soon became *de rigueur* on the celebrity party circuit to boast the ownership of such an exciting talking-point. In Manhattan society, there was 'radium roulette', and 'radioactive cocktails'; there were 'radium dances' and even a musical, *Piff! Paff! Pouff!*, in which the wonders of radium provided the plot.[20] In Paris, an exotic dancer from Illinois, Loie Fuller, had been enchanting audiences at the Folies-Bergère and at the Eiffel Tower during the 1900 Exposition, as 'the light fairy' with her extravagant use of 'the magic fluid of electrical light effects'. Having heard about the fascination of radium, she boldly wrote to Marie Curie for advice on using the new magical element in a spectacular costume which she characterised as 'butterfly wings of radium'. The amused but generous Curies let the over-imaginative dancer down gently, but the correspondence led to Fuller several times arranging private performances for the Curies at their home in Boulevard Kellermann. These events involved electricians and others trailing through the house amid the wholesale shifting of furniture, but as the unlikely friendship developed, the Curies made return visits to Fuller's home, where they met Auguste Rodin, among others of the artistic community of Paris.[21] Later, in May 1919, the patrons of Denver's famous Tabor Theatre revelled in 'The Radium Models' vaudeville act, in which women wearing luminous costumes allegedly dosed with radium posed as well-known sculptures.[22] The new element, for so long the province of academics and which promised so much for medicine, looked like being hijacked by everyone from bartenders and choreographers to celebrity 'physicians' and the US government. For whatever reasons, but probably largely the popularity of luminous materials, the prospects for the radium industry looked good:

> Radium is being produced in Denver by the Schlesinger Radium Co., using a secret process that is said to be superior to that developed by officials of the U.S. Bureau of Mines. The new company has offices and laboratory at 2001 East Colfax Avenue in the Capitol Hill residence section; the works are at 1045 Tejon Street on the west side, convenient to railroad trackage. The laboratory site was selected with the object of securing as great isolation as possible from vibrations and other disturbing influences that prevail in the manufacturing part of the city. The production of radium averages, at present, about $1,000 worth per day. The U.S. government is reported as owning 560mg of radium bromide that was manufactured in its Denver plant prior to its sale to the Pittsburg Radium Co.. [23]

Even in the consumerist USA, however, there was growing disgust in some quarters at the hysteria for the trivial use of radium, to the detriment of medicine and other more serious purposes:

> The use for radium in luminous compounds has been discouraged and for good cause. The radium so used is lost, and since the known deposits of radium ores are being rapidly exhausted, the future supply of radium for use in therapy should be gravely considered and no waste permitted at the present time. [24]

The entertainment industry in Britain did not get its hands on luminous paint in quite the extravagant American way. In August 1918, the Ministry of Munitions issued a Prohibition Order on the purchase, sale or delivery of radioactive substances and luminous bodies or ores without a permit:

> The Order applies to all radioactive substances, including actinium, radium, uranium, thorium and their disintegration products and compounds, luminous bodies in the preparation of which any radioactive substance is used, and ores from which any

> radioactive substance is obtainable, except uranium nitrate and radioactive substances which at the date of this Order form an integral part of any instrument, including instruments of precision or for time-keeping.[25]

The war, with the resultant difficulties of supply of ores, wrought great damage on the world radium industry. In France, carnotite was still obtained from Colorado, but new supplies of autunite were being exploited in Portugal and a variety of minerals had been discovered in the French colony of Madagascar. The meagre mineral resources of Cornwall were insufficient to maintain the industry in Britain. The British Radium Corporation was in receivership by 1918 and was dissolved in March 1921. A number of other companies suffered a similar fate: the International Vanadium Co. in Liverpool was dissolved in 1916; the Lumine Co. (to acquire the 'Lumine patents') went in 1923; Radioactive Waters Ltd dried up in 1922; the Radium Colour Co. faded in 1922; the Radium Development Syndicate collapsed in 1924; Radium Electro Emanations Ltd appointed a receiver in 1916; Radiumlamp Ltd went out in 1921; Radium Natural Springs Ltd was dissolved in 1921; Thomas Thorne Baker's Radium Salt Co. in Carrickfergus also ended in 1921. Seventy per cent of the radium-related companies that had been incorporated in London after 1908 collapsed during or immediately after the First World War.[26]

John MacArthur, although producing radium for the Admiralty's luminous paint needs, was fully aware that there were good arguments for avoiding unnecessary exploitation:

> While it is desirable that as much radium as possible should be available for the medical profession, and that it should not be wasted on articles of luxury, the legitimacy of its use for illuminating ships' compass cards is rather a different question. It has been suggested that mesothorium might profitably take the place of

> radium in the manufacture of luminous compounds. The radio-activity of this element is for practical purposes comparable with that of radium, but its life is measured by years, not centuries, and it could not with advantage be recovered more than once or twice.[27]

Because of the restrictions both of wartime and of the US government in placing embargoes on the export of ores, MacArthur had spent many weeks during 1918 examining alternative sources of supply. In February, he visited mine sites at Tolgarrick, Lostwithiel and South Terras, and his notebook from the period is full of cryptic comments, sketches, measurements and calculations; of South Terras, he noted:

> French company took it over about five years ago and found the pumps and engine not up to their work, partly owing to structural defects of engine and pump but also to excessive amounts of water. Main shaft – there is no ore above 30 fathom level. French Co. pumped dry – sampled but not worked. These pumps are still on the ground but require a good boiler.[28]

In May 1918, MacArthur visited Wheal Edward and Wheal Orrles, near St Just:

> Mines are about 1m. from St Just P.O., and sea about 1m. further. The old shafts were not visited though U had often been found in the tin ore. Approach short but very rough. Samples containing pitchblende were taken from boulders on the shore.[29]

MacArthur's notebook is full of shorter notes and figures relating to various sites. There were later visits to mines at Guarda in Portugal and in Spain, and his notes are interposed with incidental references to steelmaking in Sheffield, 'molybdenum is coming', and radium fertiliser – lists of quantities and initials, including someone who 'wants five tons'. His opinion of the resources of Cornwall was high. Writing for *The Mining Journal* on the perils of the wartime shipping of ore, he offered a powerful comparison:

> Much valuable shipping space might, however, have been saved had it been found possible to make use of considerable quantities of radioactive ores which are known to be in Cornwall. The records of pitchblende production in that county are admittedly imperfect, and this imperfection accounts partly for the failure to appreciate the importance of Cornwall as a producer. What it might do may be judged from the fact that one mine alone, to the records of which the writer had access, produced in four years ore which contained radium equal to 18,000 milligrammes of radium bromide, an amount comparable with the output of all the carnotite fields of America in the first four years of active production.[30]

One potential new ore development in Britain came to light during the war. A private speculator discovered a rich pitchblende lode on the Kingswood Estate at Buckfastleigh, near Ashburton in Devon. The ore was analysed at University College, London, where it was found that the samples showed a uranium oxide content of over 26% – over thirteen times richer than the ore which was being imported from America.[31] A sample of pitchblende examined in 2003 gave radiation readings approximately eighty times that of normal 'background' for the surrounding area, and entering the mine was declared inadvisable.[32] There is no evidence that this find was ever commercially developed.

In the early 1920s, perhaps responding somewhat desperately to the supply problem, the Government Laboratory in London conducted research into the feasibility of recovering radium from the luminous dials of watches, clocks and all sorts of dials and other objects. In theory, after a few years, breakdown of the zinc sulphide rendered the luminous paint ineffective, thus allowing the possible recovery of the radium, which would still have retained its potency. In practice, the paint was scraped from the luminous objects, producing a rough, random mixture of flaked paint, varnish, wax, enamel, pigments, mica, asbestos, broken glass and small pieces of metal. The process of recovery was complex, but was claimed to be easier than expected. However, this was another area in which there is no evidence that experiments ever led to radium recovery on a commercial scale.[33]

John MacArthur was one of the wider scientific community who consistently voiced concerns about reorganising society. His interest in arranging the proper reconstruction of science and industry after the war might have been followed by his further involvement in the moral and ethical questions which were becoming so urgent. However, at the beginning of March 1920, he suffered a stroke and was confined to bed. Everywhere in Europe, institutions were seeking a new and more secure future following the end of the war. The League of Nations had just been launched in London. There was plenty for it to do. Many of the great issues of the time are unresolved still. Peace in The Adriatic was one of the great themes of the day, and the carve-up of Turkey was also exercising the leader-writers. Not to mention Ireland: while the British government published its Irish Home Rule Bill, and London policemen gave up horses in favour of cars, the Mayor of Cork was assassinated and a Dublin magistrate was dragged from a tramcar and killed. By the end of March, there had been 29 political murders in Ireland in three months, and the British government made the infamous decision to send 800 'Black and Tans' across the Irish Sea. The *Glasgow Herald* covered all the big international stories during March, but also dealt with the horrors of post-war rising prices: 'the rise of threepence in condensed milk by the Nestlé and Anglo-Swiss Condensed Milk Co.'; 'the ninepenny loaf' which was referred to in Parliament, angering Scots, who had always been used, almost as a matter of course, to paying higher prices than the English – 'it is eighteen months since the controlled price in Scotland was raised to 9½*d* and even higher in remote areas'. On 17 March the newspaper reported the death of John MacArthur. The after-effects of the stroke had become complicated, and he died quite suddenly at home at Knowe Terrace on the evening of Tuesday 16 March 1920, at the age of sixty-three. The British radium industry seemed to be collapsing. MacArthur's obituary notice in the science journal *Nature* gave a representative tribute, referring to his earlier international success in the field of gold and silver extraction:

> By the death of Mr J.S. MacArthur on March 16th, industrial chemistry has lost a notable exponent. Mr MacArthur's name will always be remembered in connection with the MacArthur-Forrest patent for the extraction of gold from its ores by means of cyanide. It is given to few men to discover a process which has had such a far-reaching effect in almost every branch of civilised life. After this work was completed, Mr MacArthur engaged in many commercial ventures in connection with chemistry and mining, but, with the possible exception of his last, none of them seemed to possess the elements of permanent success. This was the extraction of radium from its ores, which he carried out first of all in Cheshire, and then practically on the shores of Loch Lomond, in order to avail himself of the purest possible water. He was proud of his works there, and delighted to feel that he was able to carry on his work in the midst of such beautiful surroundings. [34]

Under the terms of John MacArthur's will, the Loch Lomond Radium Works was disposed of in various proportions between his brother Charles Stewart MacArthur, his nephew John Smallwood MacArthur, and William and George Dempster. John Smallwood MacArthur gave up the post of company secretary and went to work and live initially in Cleveland, Ohio, before settling in Toronto. The Dempsters established an office in Manchester, probably for marketing purposes, and continued the business, for a time, along the lines established by MacArthur. However, they were unable to carry on successfully, and in April 1922 the decision was taken to close the business.

George Dempster left Loch Lomond-side for South Terras in Cornwall, taking several tons of ores and residues with him. La Société industrielle du radium Ltd had kept the mine inactive, as the cost of transporting the rather poor ore to France was prohibitive. However, at the time of Dempster's arrival a small refining plant was built, the first of its kind in Cornwall, and radium bromide, radium–barium chloride and uranium, polonium and actinium salts (and a radioactive fertiliser) were produced. Discussions opened over a proposed amalgamation between La Société industrielle du radium and Radium Ore Mines Ltd, which owned the nearby Tolgarrick Mine, with a view to forming English United Radium as a new joint company. Production at South Terras continued until 1927, when a dispute with the Boconnoc Estate induced the collapse of La Société industrielle du radium. However, following the closure in 1928, a German–Swiss company named British and General Radium Corporation, took over, building a laboratory at Trevarthian Road in St Austell, and planned to extend the ore lode.

Dempster became very ill and returned to Scotland in 1930, where he lived as an invalid until his death from tuberculosis in 1934. The hoped-for expansion at South Terras never took place and the mine, once over-optimistically thought to be the largest known deposit of uranium and radium in the world, was abandoned in 1931.[35] The estate tried unsuccessfully for many years to re-establish some form of working at South Terras, including the development of a spa in 1937, but South Terras was doomed, the victim of years of under-funding and under-development. It is an interesting observation from local gravestones that many of the workers at South Terras lived to a ripe old age, suggesting better conditions than usual; pitchblende was probably more profitable than other local ores, and digging at South Terras only took place in summer months, when the mine was dry. Perhaps the better than average conditions tended to lengthen the workers' lifespan more than radioactivity tended to shorten it.

Regardless of its small place in the story of radium, South Terras enjoyed the attention during its working life of several French chemists, including two of Marie Curie's personal assistants. Jacques Danne, who had also been the Chief Editor of the very early journal *Le Radium*, worked there in 1913 when La Société industrielle du radium began its operations. He was later joined by Marcel Pochon, a young chemist born in Versailles of Swiss parentage, who had graduated from the rue Lhomond in Paris in 1907. He worked at South Terras for several years until its final closure, when he went on to notable success in radium production in Canada.

Despite all the bureaucratic and commercial angst in the USA, and the rather more restrained concern in Britain and other European countries, the development of entirely new and unexpected sources of ore was to be momentous. In August 1915 a British geologist, Robert Sharp, working for the Union minière du Haut-Katanga, accidentally stumbled across heavy rocks laced by green and yellow streaks near Elizabethville (now Lubumbashi) in the Haut-Katanga area of the south-eastern Belgian Congo (now known, after many years of bloody turmoil, and a period as Zaïre, as the Democratic Republic of Congo). Sharp identified the mineral as pitchblende – and it had been lying on the surface! Laboratory analyses and experimental processing at Elizabethville and in Belgium indicated exceptional ore richness. Despite the extremely high grade of these deposits (up to 68 per cent uranium oxide, the highest ever found), wartime pressures ensured that exploitation of the equally rich copper, tin, cobalt and manganese resources would receive priority. Sharp went back to Chinkolobwe and marked out the claim with a simple wooden pole with a zinc

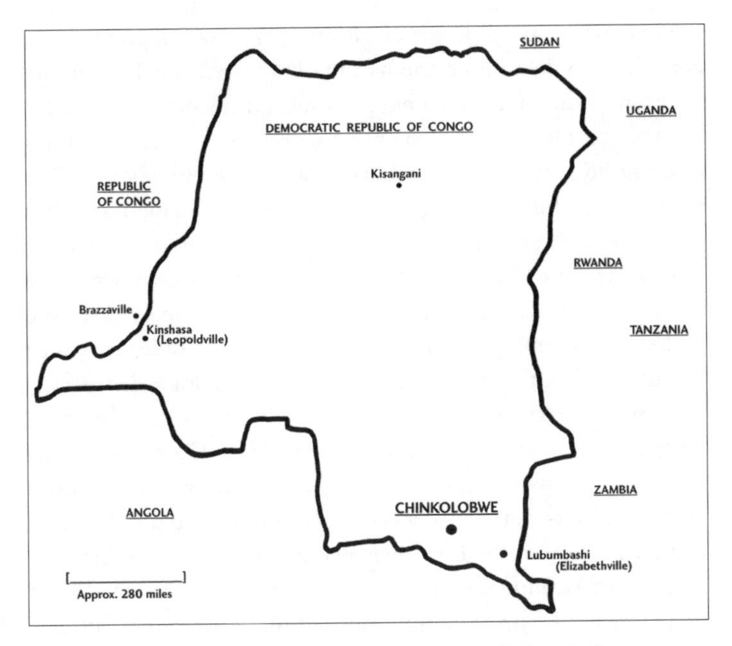

Map showing the location of Chinkolobwe, (Belgian) Congo.

plate bearing one word – Radium. He went off to war in Europe, hardly expecting that years of inaction would follow; however, his casual kick of the boot would eventually launch the world's first nuclear weapons.[36]

The Belgian mining trust Union minière du Haut-Katanga and its parent, the Société générale de Belgique, controlled everything that moved in Haut-Katanga, and the new discovery remained secret until after the First World War. When Robert Sharp returned in 1919, he realised that nothing had been done to register the Chinkolobwe Mine. He set the wheels in motion, and the following year an extraction plant was constructed at Sint-Jozef-Olen, near Antwerp, by a new partner, the Société métallurgique de Hoboken, which had already undertaken other mineral extraction processing.[37] Marie Curie was appointed as a consultant to the project, and had an office at the site. The remote location of the mine was a big physical and financial problem, as was the complex nature of the ores themselves – variable mixtures of carnotite, pitchblende, torbernite and uranium silicates. This would be no high-tech business, however. Local tribesmen were supplied with picks, shovels and hammers and earned their pittance by scrabbling across the open-cast surfaces, hand-picking the rocks and loading them on buffalo carts for the trek to the railhead for shipping to Antwerp. In later years, in order to reduce the high cost of transporting crude ores over long distances, a small preliminary purification plant was erected at Chinkolobwe, with the more intricate stages conducted at Olen. Processing took place in two major stages: firstly of sixty crystallisations in enamelled cast-iron pans of successively diminishing size; secondly, further fractionations in smaller quartz basins, during which the amount of radium was increased from 0.05 per cent to about 98 per cent. The first radium produced using the enormous new sources of ore in Africa appeared in 1922; the new facility was spectacularly successful and 12g of radium was produced in the first year. The radium bromide was delivered to Union minière's headquarters in Brussels, where it was prepared in applicators for commercial and therapeutic uses, and from where it was marketed under the name Radium Belge. The company produced a huge range of radium applicators and accessories, and all radium products were accompanied by a certificate of authenticity. This guaranteed calibration against a tube of radium chloride extracted from minerals at Chinkolobwe which had been accepted by the International Standards Commission under the supervision of Marie Curie and Lord Rutherford.

The writing was on the wall for the earlier principal world sources of radium ores in Colorado, Utah and Portugal. It has been estimated that Olen produced around half (4.5kg) of the total world production of radium, up to about 1930.[38]

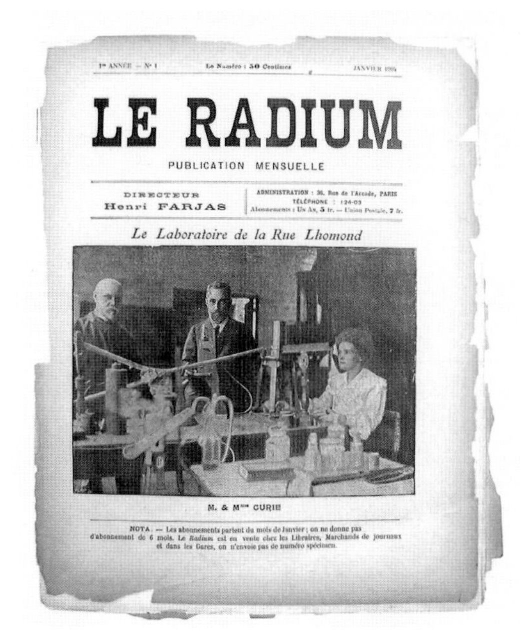

1re ANNÉE — N° 1 Le Numéro : 50 Centimes JANVIER 1904

LE RADIUM

PUBLICATION MENSUELLE

DIRECTEUR
Henri FARJAS

ADMINISTRATION : 36, Rue de l'Arcade, PARIS
TÉLÉPHONE : 124-03
Abonnements : Un An, 5 fr. — Union Postale, 7 fr.

Le Laboratoire de la Rue Lhomond

M. & Mme CURIE

NOTA. — Les abonnements partent du mois de Janvier ; on ne donne pas d'abonnement de 6 mois. Le *Radium* est en vente chez les Libraires, Marchands de journaux et dans les Gares, on n'envoie pas de numéro spécimen.

1 First edition of *Le Radium* showing the Curies in the laboratory in rue Lhomond, Paris. (Courtesy of Dr Paul Frame, Oak Ridge Associated Universities)

2 Sir William Crookes in 1900.

3 Typical method of radium production, *c.*1905. (*Le Radium*)

4 Industrial-scale radium production, Nogent-sur-Marne, *c.*1910.

5 The Radium Palace Hotel, St Joachimsthal. (Wellcome Library, London)

RADIUM-CURE ESTABLISHMENT

ST. JOACHIMSTHAL

Near KARLSBAD (BOHEMIA).

OPENING, MAY 15, 1912

FIRST-CLASS KURHOTEL

. . . 300 ROOMS . . .

Prospectus and all information
to be had from the Management,
"RADIUMKURHAUS,"
St. Joachimsthal, Bohemia
(Austria).

6 Advertisement for the opening of the Radium Palace Hotel.

7 John Stewart MacArthur. (Courtesy of the Master and Fellows of Balliol College, Oxford)

8 Sir William Ramsay. (Originally published in *A Life of Sir William Ramsay* by Morris W. Travers, Edward Arnold, 1956)

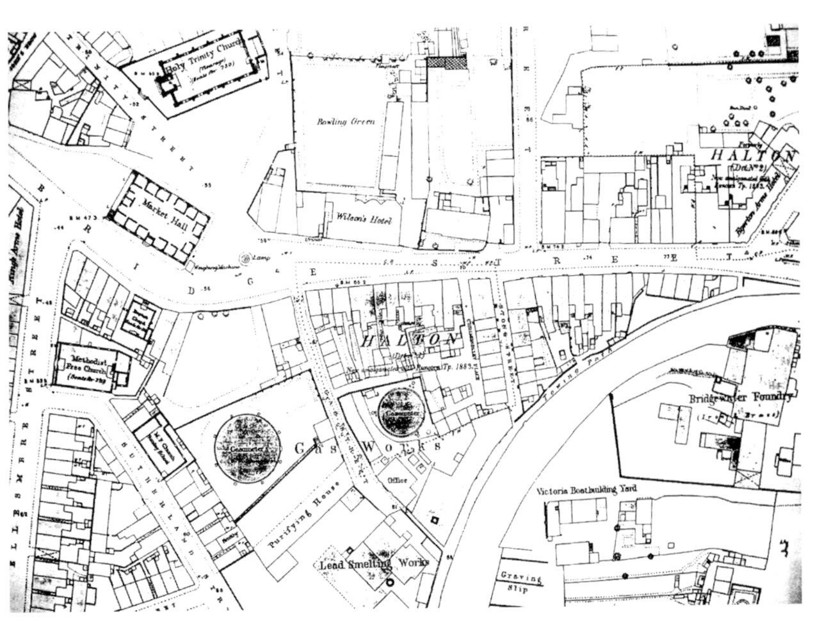

9 Ordnance Survey map of 1876 showing the site of the former Radium Works on the bank of the Bridgewater Canal, at Halton near Runcorn in Cheshire. (Reproduced by permission of the Trustees of the National Library of Scotland)

10 Housing on the site of the former Radium Works at Halton.

11 The Loch Lomond Radium Works, Balloch, *c.*1928. (Courtesy of West Dunbartonshire Libraries)

12 Site of the former Loch Lomond Radium Works, 2004.

13 Evidence of the fertiliser's efficacy on hyacinths, shown by two young relatives. (See advertisement on p.13)

14 Staff at South Terras Mine, Cornwall, *c.*1930; Marcel Pochon, in cap, is seated in the middle of the second row. (Courtesy of Gill Pearce)

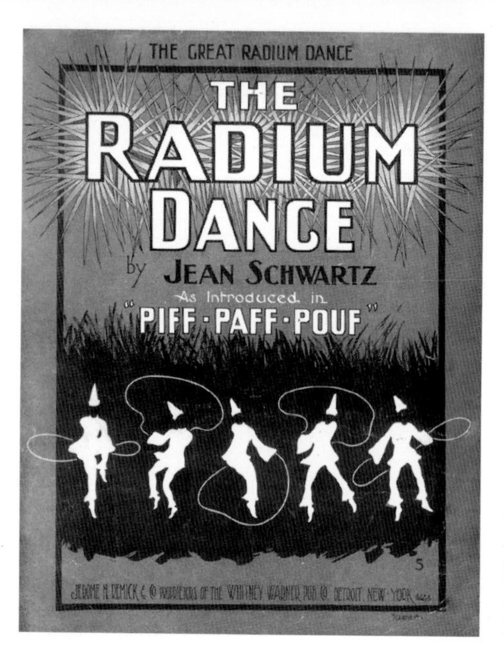

15 Sheet music cover for 'The Radium Dance', by Jean Schwartz, for the musical *Piff, Paff, Pouf*, 1904. (Courtesy of Special Collections Library, Duke University, Durham, North Carolina)

16 Sabin von Sochocky. (Courtesy of US National Archives & Records Administration, Chicago)

17 A 1917 advertisement for Undark luminous products.

18 Workers at the Radium Dial Co., Ottawa, Illinois, 1925. (Courtesy of US National Archives & Records Administration, Chicago)

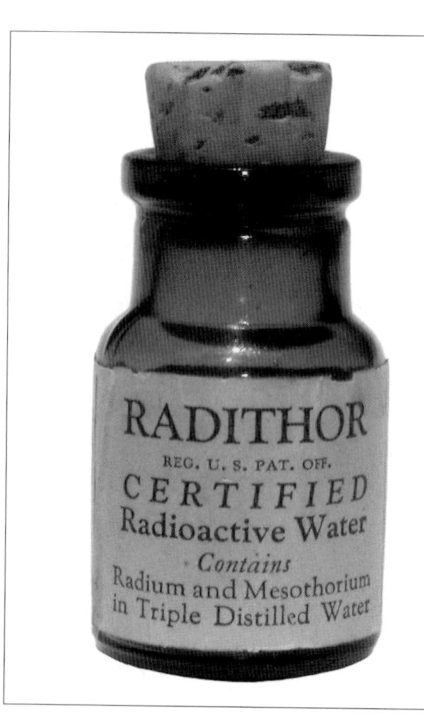

19 The US Radium Corporation dial-painting studio, West Orange, New Jersey. (Courtesy of US National Archives & Records Administration, Chicago)

20 'Radithor' radium-water, sold by William Bailey. (Courtesy of Dr Paul Frame, Oak Ridge Associated Universities)

21 The Olen Radium refinery near Antwerp in Belgium. (Published *c.*1928 by L'union minière du Haut-Katanga)

22 Radium at Olen, photographed by its own radiation. (Published *c.*1928 by L'union minière du Haut-Katanga)

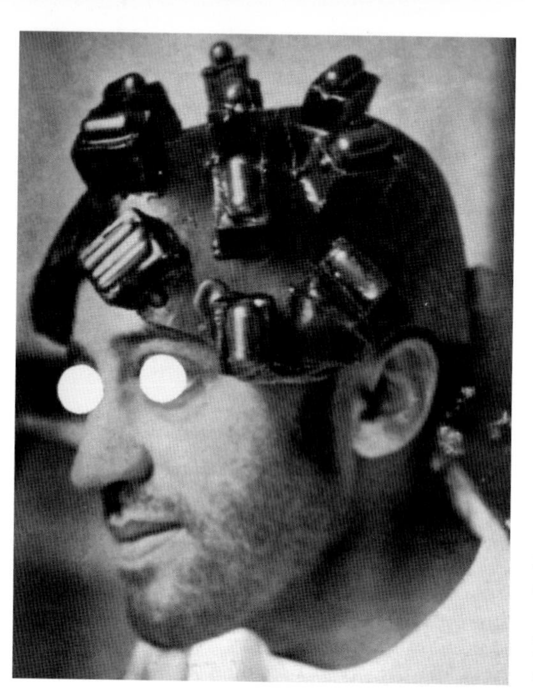

23 One of a wide range of medical radium applicators – in this case for treating cerebral tumours. (Published *c.*1928 by L'union minière du Haut-Katanga)

24 Gilbert LaBine at Great Bear Lake, May 1930. (Courtesy of the National Archives of Canada)

25 Aerial view of Port Radium in the 1930s. (Courtesy of the National Archives of Canada)

26 Port Radium in the 1930s. (Courtesy of the National Archives of Canada)

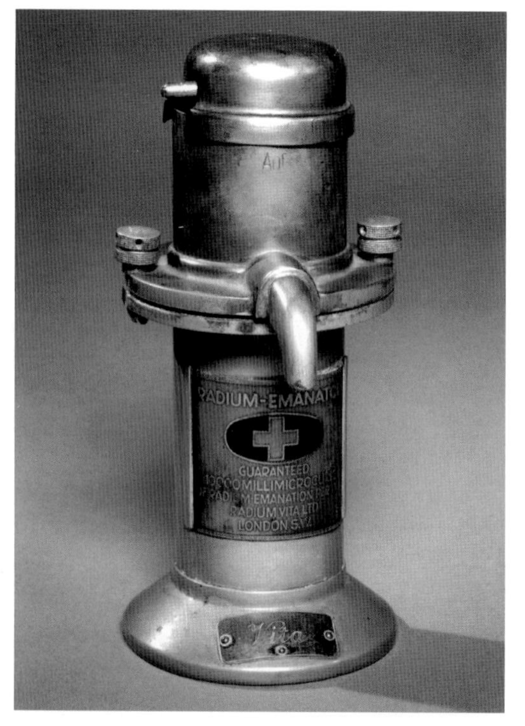

27 Radium refinery at Port Hope, Ontario, 1938. (Courtesy of the National Archives of Canada)

28 Radium emanator produced in London. (Courtesy of the National Radiological Protection Board)

29 Radium Ore 'Revigator' produced in San Francisco. (Courtesy of Dr Paul Frame, Oak Ridge Associated Universities)

30 The 'Favorite' Swiss-made DIY luminising kit. (Courtesy of Dr Paul Frame, Oak Ridge Associated Universities)

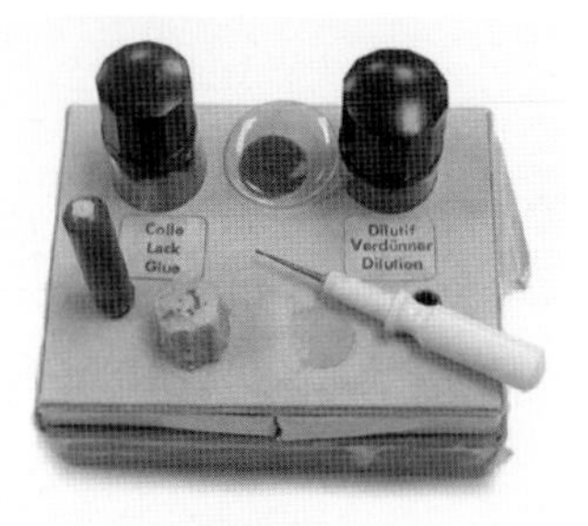

31 *Above:* Contents of the 'Favorite' luminising kit. (Courtesy of Dr Paul Frame, Oak Ridge Associated Universities)

32 *Right:* The Denver Radium Service Emanation Bath. (Courtesy of Dr Paul Frame, Oak Ridge Associated Universities)

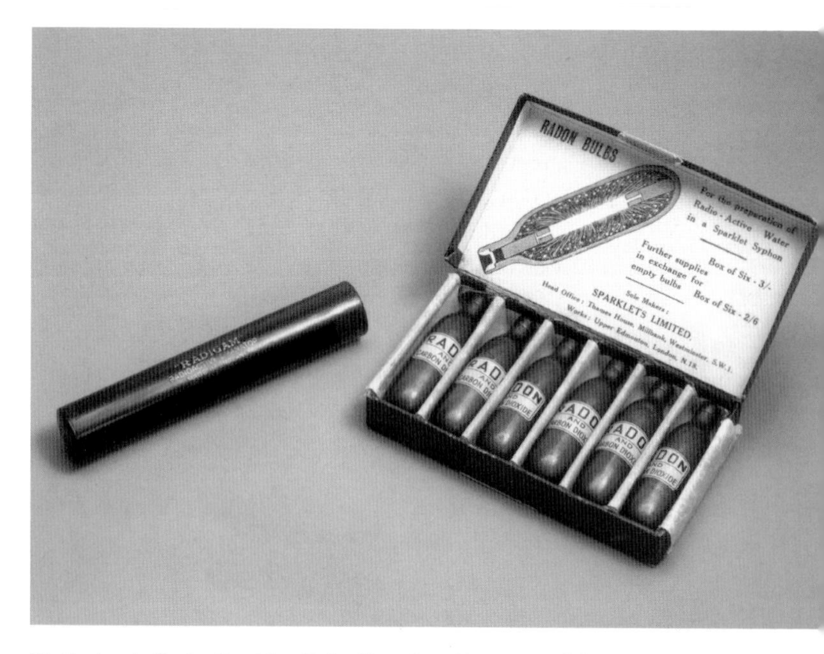

33 Radon bulbs by Sparklets Ltd of London. (Courtesy of the National Radiological Protection Board)

The cost of Belgian production was low (perhaps artificially), and soon the Standard Chemical Co. of Pittsburgh abandoned radium production, and other plants followed; by the summer of 1923 only the US Radium Corporation remained to stave off a collapse that would happen in 1925. American pride was salved by Standard Chemical and the Radium Co. of Colorado acting briefly as technical advisers and marketing agents to the Belgian company.[39] The US radium industry began to pick up again when it concentrated once more on extracting vanadium for the steel industry. In Belgium, Olen's price kept falling, but Union minière consistently refused to publish its production figures, and its archive has never been opened to scrutiny (a level of secrecy intensified after the Second World War, when the USA acquired much of the uranium for its vast nuclear bomb-making activities from Chinkolobwe). Production at Olen was noted for its consistent output of about 3g per month; radium was produced with clockwork efficiency – for every 3 tons of hand-picked ore from Chinkolobwe, 1g of radium resulted. Such efficiency ensured that huge reserves of yellowing residues built up at the site – residues that contained very high ratios of uranium, powerful enough, it soon became known, to obliterate half the world. When pitchblende sources were developed in Arctic Canada in the mid-1930s, the Belgian and Canadian factories together drove the price of radium further downwards, ensuring a world duopoly just as the demand was beginning to change from radium for medicine to uranium for much less humanitarian purposes.

For the moment, the radium story pauses during the First World War, with the future of the industry in considerable doubt, and with no-one able to foresee impending events. Considerable scepticism on the subject was still common. In December 1919, as the twentieth anniversary of the discovery of radium arrived, a letter to *The Times* under the heading 'The Waste of Mechanical Power' tried to put the whole radium hysteria in perspective:

It is to be hoped that the wasters of our coal deposits will not be further encouraged by the frequent assertions that we may be on the eve of an unlimited supply of energy from radium or other sources. In any practical sense there is about the same chance of our being able to import coal from the moon. It is a striking and wonderful fact that we happen to live in an age which will in a few brief generations have exhausted the natural deposits – coal, ores, oil – which have required aeons of geological time to accumulate. 200 years ago the human race knew practically nothing of these powerful means for the mastery of physical conditions of life; 200 years hence they will possibly be looking back on an era, almost a moment in the history of the race, that irrevocably exhausted them all.[40]

EIGHT

'INFUSED WITH THE HIGHER VIBRATIONS'

Even twenty years after the discovery of radium there was still little understanding of the likely dangers or of the diagnosis of radiation damage. As early as 1903, however, there had been serious concern expressed about the dangers of working with X-rays. Clarence Dally, one of Thomas Edison's assistants, was an early and well-publicised victim:

> That loss of sight, cancerous disease and even death may come to him who is constantly exposed to or inexperienced in the use of Roentgen rays has been demonstrated in a pitiable manner in the laboratory of Thomas A. Edison at Orange, N.J. Clarence Dally, an assistant to the 'Wizard of Menlo Park', has contributed an arm and a hand to this demonstration, while Mr Edison himself suffers from the disturbed focus of one of his eyes through experiments with this mysterious light in an endeavour to find for it some commercial utility.[1]

Edison was appalled by what had happened to his assistant, and promptly gave up any further interest in such work:

> Don't talk to me about X-rays. I am afraid of them. I stopped experimenting with them two years ago, when I came near to losing my eyesight and Dally, my assistant, practically lost the use of both of his arms. I am afraid of radium and polonium too, and I don't want to monkey with them. [2]

Ironically, Dally was treated with further doses of X-rays in a futile attempt to undo what damage he had suffered, and further amputation was undertaken before he died the following year at the age of thirty-nine. As late as 1914, knowledge of X-rays was still being acquired by trial and error. One professional journal opened with the note that 'we have to deplore once more the sacrifice of a radiologist, the victim of his art.'[3]

With much less serious professional objection to radium than was being attached to X-rays, there was little likelihood that the public was about to recognise any problem. There was something exciting about the idea that such a sober, 'scientific' substance from the higher reaches of physics, chemistry and medicine could also produce the fun that luminous paint offered, and that was enough to maintain public enthusiasm. The new fad, however 'magical', appeared in truth to be just like one of the many patent remedies, several of which were more easily recognised as dangerous. There were nevertheless endless melodramatic stories of the properties of radium which – whether true or fanciful – were probably inevitable given the public's perception of the 'magic' of radiation. It seemed unlikely that the public would ignore the substance that could 'cure cancer and blindness, determine the sex of unborn children and turn the skin of Negroes white'. The public was bombarded with far-fetched comparisons of the amazing energy produced by a gram of radium and a ton of coal; another pundit announced that a gram of radium could raise 500 tons a mile high, while an ounce could drive a 35hp vehicle at 30mph round the world. Further bizarre demands on the imagination that had increasingly small amounts of radium transfer the entire British naval fleet to the summit of Mont Blanc seemed to deter no-one.[4]

Although notorious instances of death and injury in the late 1920s would secure public notice with a vengeance, the silent danger to the public undoubtedly came from the galaxy of products containing radium during the 1920s and '30s – many of which remained on sale for decades. There were lotions, potions, tablets, inhalations, injections and suppositories for insertion in every bodily orifice for ailments known and undreamed-of; and douches, pads, belts and 'applicators' for the improvement of every part of the body, not excluding that most unreachable of extremities – the human spirit. You could use Radiogen, a German radioactive toothpaste claiming to release radon as you brushed, promoting both hygiene and digestion.[5] Your efforts might successfully brush away the detritus of a German radium chocolate bar sold as a 'rejuvenator'; you might pep up your gin with tonic-water produced by a radium Sparklets bulb, or use radioactive digestive mixtures, face creams or even a radium-based contraceptive jelly still being sold in America in the mid-1950s. One of the boldest ventures was by Associated Radium Chemists of New York (whose president was one William Bailey, of whom more later) who simply sold tins of 'Arium' radium tablets ('two with a glass of water before or after each meal'); this nostrum was ordered to be destroyed following a court investigation. Many more similar preparations and devices were entirely frivolous or fraudulent, but there were endless products which claimed specific medical attributes; as late as 1929 a European pharmacopoeia listed eighty patent medicines claiming radium as the principal efficacious constituent.[6] Epitomising the go-ahead companies devising new radium products for sale to the public was the Denver Radium Service in Colorado; its range included:

Suppositories for rectal, vaginal and urethral insertion;
Radium ointments for chest colds, skin irritations, wounds and sores;

> 'Radiumactive Vitalizer' for producing radiumized, gas-charged health water;
> Ophthalmic solution for hay fevers and cataracts;
> 'Raditone' tonic tablets containing radium plus gland extracts and enzymes for relief of constipation, stimulation of mental processes, and increased sexual vigour;
> 'Radium Emanation Bath Salts';
> 'Narada Radium Preparations' – a line of beauty preparations including lemon massage cream, anti-wrinkle cream, muscle-toning oil, masque and bleach pack, rouge, and mascara for the smart American girl and woman. [7]

When the fad for radon spas and radon inhalation in hospitals was in full swing, entrepreneurs and practitioners attempted to bottle the radioactive waters for the benefit of those who were unable to visit the natural springs. However, they found that radon in water was a fickle thing; as the gas either decayed or escaped to the atmosphere very quickly, the efficacy was lost to the multitude of potential home users. Money was also to be made at the entertainment end of the business by selling cocktails and drinks that glowed in the dark. Finally, the goal was to sell to the public devices that contained a radium source which could be used to add radon to drinking-water, beer, milk, etc. These devices, known as 'emanators', became enormously popular on both sides of the Atlantic. They took the form of a pottery crock or jug made partially of radioactive ore embedded in the clay, or a metal container incorporating a radium source crudely sealed in asbestos or cement; the liquid to be irradiated was left within the jug for an appropriate period before being drawn off as required by a tap. If the drink was required to be entertainingly luminous as well as allegedly stimulating, the emanator often contained an activator such as zinc sulphide. After it was realised that radon ingested in this manner was rapidly expelled from the body, some 'advisors' encouraged 'sipping séances', at which patients consumed many small amounts of charged water at short intervals in order to 'keep their levels up'.[8] The popularity of such emanators and the consequent radioactive material casually ingested meant that they were perhaps the most damaging products of the radium age.

In Britain, emanators were produced by Radium Utilities Ltd, and by the Radium-Vita company (boasting offices in London, Paris and New York), which also manufactured a range of face creams and radium ointments, produced in London until 1954. Another small, tube-like emanator about 12cm in length, the 'Radigam', was produced by Sparklets Ltd, whose showrooms were at Millbank in the heart of Westminster. They were best known for their Sparklets Radon Bulbs; in boxes of six, of two strengths, costing either 3*s* or 7*s* 6*d*, they were supplied from a factory in Edmonton, North London. They were used 'for the preparation of Radio-Active Water in a Sparklet Syphon' (the siphon cost an extra 6*s* 9*d*). Sparklets ('a palatable drinking water of vastly higher radioactivity than any existing spa') were marketed under the name 'Radium-Spa'. The No.1 Tonic-size bulb contained 1,050 maché units (certified by the National Physical Laboratory), compared to only 201 maché units available at the world-famous Gastein Spa in Austria, which claimed the world's highest level of radioactivity in natural water. In the USA, the most popular model of emanator was the Radium Ore Revigator, manufactured in San Francisco, and sold in hundreds of thousands. This was lined with a porous radium ore, and when filled with water and left to stand overnight, produced potent radon water each morning:

> The Millions of tiny rays that are continuously given off by this ore penetrate the water and form this great Health Element – Radioactivity. All the next day the family is provided with two gallons of real, healthful radioactive water ... Nature's way to health.[9]

The idea was that ordinary drinking water had a 'basic fault', but the company was smart enough not to make direct claims of miraculous cures, but aggressively published such naïve claims offered by their customers in the form of glowing testimonials. Thus developed the wondrous ability of the Radium Ore Revigator to conquer arthritis, 'lost manhood', nervousness, neuritis, and everything else from piles to the inevitable 'female troubles'. As the American Medical Association observed:

> While not directly recommending the Revigator for sexual impotency and glandular deficiencies, they mention that radium emanation 'increases sexual activity' and 'stimulates ... the entire glandular system' – and allow the sexual neurasthenic to draw his own conclusions.[10]

The company also employed 'independent' medical practitioners, such as 'Dr Donald Donovan' of Los Angeles, to puff their product:

> Your Radium Charging Jar can be appropriately called 'The Jar of Life', as water placed therein becomes infused with the higher vibrations, and constitutes a rational and perfect method to retain and regain health and prolong life. [11]

According to the American Medical Association, in order to obtain the equivalent of the minimum adequate doses of emanation, a user of the Revigator would have to drink the entire contents of 60 to 120 Revigator jars every day. The devices made by such companies attracted a great deal of concern and notoriety, and the AMA regularly publicised their unfounded claims:

> As is commonly the case with latter-day pseudo-medicine having large financial resources behind it, the Revigator concern put forward an hypothesis for which there is no foundation and

> proceeds to build its claims upon it. The thesis in this case is that the drinking water of today has a 'basic fault'. It is 'denatured'; more than that, it is 'tired or wilted water'. To quote from the Revigator advertising, 'More illness is caused by improper water than any other reason and largely because radioactivity is lost from our daily supply of drinking water.'[12]

Other emanators available in America were the Lifetime Radium Vitalizer Jar, an aluminium jar with radium ore in the base, produced in Chicago and Denver, and the Torbena Jar made in Allentown, Pennsylvania, containing torbernite mined in New Mexico. Other models designed to be placed under the water were the Radium Life Emanator of Los Angeles; the Thomas Radioactive Cone, and the Zimmer Radon Generator; some devices were essentially 'cottage industry' productions – anonymous crocks often supplied to hotel guests in their rooms. The Radio-Rem Emanator, produced by a New York pharmaceutical company, consisted of eight small bottles each containing a ceramic rod which released radon into the water over a four-day period; by the time the eighth bottle had been used, the first would have been recharged.[13]

A trawl through patent applications reveals the imaginative range of products and devices which were prepared in Britain for modification by the addition of radium. For a really healthy home life, you and your family could have a radium ventilation system. A patent of 1909 proposed a system which forced air through a pair of perforated plates containing radioactive material, 'to charge the air with their emanations'.[14] And radioactive clothing could also be obtained quite simply by impregnating an otherwise ordinary fabric:

> Wearing-apparel is rendered radioactive by treating the dried fabric with a solution of carbonate or bicarbonate of soda or potash, or a sulphate which will convert a soluble radioactive salt into an

> insoluble substance, after which the fabric is dried and immersed in a solution of radioactive salt such as bromide or chloride of radium or a solution of a salt of thorium or uranium. [15]

In a London advertisement of June 1929, the *Radio Times* came up with something a bit unusual – and dangerous. Radioactive corsets would 'overcome the most obstinate rheumatism – a proven and acknowledged remedy. Certified by the National Physical Laboratory, Teddington.' A snip at 30*s*, certificate included. A few years later another *Radio Times* advertisement announced important news for those with hair problems. A new scientific discovery promised that water from the radioactive spring at the spa town of Matlock Bath contained secret growth-stimulating ingredients known only to Frederick Godfrey, 'the well-known British Hair Specialist, whose discovery in relation to Hair Growth is one of the most important scientific achievements of recent times.'[16] In exchange for two 1½*d* stamps ('value 3*d*', added the advertisement, trying to be helpful to the gullible) the bold Godfrey would send a bottle of the radioactive hair tonic, a book on hair growth, and a personal letter about your own particular condition. Elsewhere, for the hard of hearing, was advertised Radium Ear, a device that was intended to be hung over the outer ear, with a metal tube for insertion into the inner ear canal; this device contained the magic ingredient 'Hearium'....[17]

A whole range of mysterious-sounding products came from Radium Electric Ltd of Haywards Heath in Sussex. With a name plucked from the realms of pseudo-science, the 'Q-Ray Electro-Radioactive Dry Compress' was described as 'a relief for rheumatism, sciatica, and invaluable for pneumonia and bronchial complaints'. The Q-Ray system ('the subject of meritorious statements in the House of Commons' according to the brochure) comprised a series of canvas mats with sealed pockets containing pitchblende ore. This seemed similar to very many such products available on both sides of the Atlantic for a wide range of medical and quasi-medical conditions. (One European version, the Radiumchema Compress, was manufactured at St Joachimsthal and was guaranteed by no less an authority than the Institut du Radium at the Laboratoire Curie in Paris.) However, the Q-Ray sported an electric heating element, and with the blanket wrapped around the appropriate part of the body, was plugged into the electric mains for the recommended period of time. The instructions did say:

> The Q-Ray Compress is not claimed to be a specific or 'Cure-All'. It has now been in public use for many years, and its efficacy and safety are beyond question.

They further insisted that, 'it is an electro-medical (radioactive) appliance for the treatment of painful conditions'. Yet again the amenable National Physical Laboratory at Teddington certified the radioactivity, and the instructions were accompanied by testimonials from an Austrian professor. Tests conducted on one of these 1930s devices found in a house in 1990 revealed that the radiation surface dose equivalent was greater than 100 microsieverts per hour[18]; today's recommended maximum dose equivalent for members of the public is 300 microsieverts per *year*. The helpful instructions did advise against going to sleep with the compress on the body and the current switched on, as 'after the current is turned off, the Radioactive Rays continue to emanate'.

Many of these and similar devices, although produced in the 1920s and '30s, have remained in circulation, often abandoned in cupboards or in attic trunks. Even the problem of the many remaining radium-luminous watches is one that requires careful handling. Surprisingly, there has been a continuing small-scale industry producing such problem items (as distinct from devices such as smoke alarms and modern luminous devices, which contain carefully modulated low-level modern isotopes, and which are wholly safe in use so long as their disposal follows appropriate procedures). In the 1960s, for example, a small-town garage in Illinois produced the Gra-Maze Uranium Comforter, a quilted bag not unlike the Q-Ray, which described itself as 'your own health mine in miniature'; this commercial rip-off was quickly put out of business by the authorities. In the 1980s there appeared from Japan the 'Endless' refrigerator-freezer deodoriser, which used thorium, with a 10-billion-year half-life, to destroy odours; this was presumably done by ionising the air – a technique employed in some early measuring devices. While banned in the USA, it is apparently popular in Japan. The same period saw the introduction of the 'Nico-Clean', a small metal plate coated with low-grade uranium ore which, when slipped into a packet of cigarettes, claimed to reduce the harmful effects of nicotine and tar.[19]

Without doubt, it was bottled radium-water that did most damage, and it was the hysteria that its use generated that led to its eventual recognition as the first indicator of just how dangerous the uncontrolled abuse of radium could become. There were numerous commercially produced bottled radium-waters, some produced in Britain and others imported from the USA and Europe. The use of radium-water appealed to those who advocated 'mild radium therapy' – the idea they peddled was that even if radium might in some way be dangerous, as some doctors were insisting, its use in small amounts was assuredly beneficial. Several brands were, inevitably, discovered to have no radium content whatsoever; one such was 'Radol' which, despite its claims of the efficacy of its radium content against cancer, contained only quinine sulphate and alcohol (nevertheless producing a helpful blue fluorescence).[20] The users of these anodyne nostrums were perhaps the lucky ones, however. Other consumers were less fortunate – they were the ones who got what they paid for. Perhaps it was to the ironic benefit of society as a whole that an early victim happened to be a wealthy American athlete-playboy industrialist, whose hideous death at the hands of a crooked charlatan hit the headlines and was followed by sustained publicity exposing the potential dangers of such casual 'treatments'. As it happened, the unqualified 'Doctor' William Bailey, the serial fraudster who perpetrated the scandal, came from the opposite end of the American social spectrum from his victim; in their different ways, they each represented a diametrically opposed personification of the great American Dream; for one of the two, the dream turned to nightmare.

William Bailey had been brought up in a tough, poor Boston neighbourhood in the 1890s and, although he had succeeded in gaining entrance to Harvard in 1903, he dropped out unqualified after two years and moved to work in a variety of commercial import–export businesses in New York. He drilled for oil in Russia, fraudulently sold patent medicines, including 'Las-I-Go' For Superb Manhood (main constituent strychnine), and established a car-sales swindle using a company named the Carnegie Engineering Corporation (with the deliberate intention of associating it in the public's perception with Andrew Carnegie's world-renowned steel company). When investigators raided the Michigan factory of the supposedly million-dollar company, it was found to be an abandoned saw-mill containing a box of rusty tools. The victims of this particular racket paid for cars that had never existed; this time, Bailey was briefly imprisoned for fraud. He was a chronic serial charlatan who after the First World War became obsessed with the prospects for medical deception. In particular, he seems to have developed a delusion of himself as some kind of 'great benefactor' who would bring to the public the wondrous advantages of radium. After his company, Associated Radium Chemists, was closed by the authorities for fraudulent trading, he set up the Thorone Co., whose star cure-all product against sexual impotence and other glandular and metabolic problems was claimed to be 250 times more radioactive than radium. He also spawned the American

Endocrine Laboratories, which conned gullible hypochondriacs out of $1,000 for his 'Radiendocrinator' which, when placed over the endocrine glands ('which have so masterful a control over life and bodily health') was claimed to cure everything in the medical lexicon, from acidosis to wrinkles, via constipation, flatulence, gout, poor memory, rickets and much, much more. Men were instructed to 'place the Radiendocrinator in the adapter ... wear like any athletic strap ... under the scrotum as it should be. Wear at night. Radiate as directed.'[21] The American Medical Association described the Radiendocrinator as 'a high-priced piece of hokum'.[22]

In 1925 Bailey established the Bailey Radium Laboratories at East Orange, New Jersey, where he devised his 'Radithor' brand of radium-water as a 'harmless cure' for over 160 ailments including, inevitably, sexual impotence; there would certainly be something for everyone. Radithor ('certified containing radium and mesothorium') was guaranteed 'harmless in every respect', and could be purchased in cases of not less than thirty half-ounce bottles for $30. Brazenly (for the man who was now desperately off-loading Arium and Thorone tablets, and lowering his price for the Radiendocrinator), Bailey insisted in his publicity material that 'all former methods of injections, emanation machines, radium ore jars, tablets, etc., have been largely discarded'. [23]

Bailey described his secret potion as 'the fullest achievement in internal radioactive treatment'; it is calculated to have contained more than 1 microcurie each of radium-226 and radium-228 (mesothorium).[24] In puffing the virtues of Radithor one newspaper reported, '"Gimme a gamma" is the cry of prematurely old humanity in search of rejuvenation.'[25] This was a period during which there was popular interest in romantic ideas of eternal youth and 'clinical renewal' of the body and spirit, and Bailey is thought to have sold 100,000 bottles every year worldwide for five years before he was stopped. 'Radithor' was a killer, and many of his customers suffered horribly.

Eben McBurney Byers was the heir to a Pittsburgh foundry company, and a millionaire socialite and famous amateur golf champion with luxurious homes in Pittsburgh, Long Island and South Carolina. Following failed efforts by mainstream medical intervention successfully to treat an arm injury and related general lassitude, he was recommended the new wonder treatment Radithor in early 1928. As a middle-aged man anxious to regain his lost vigour, Byers rejoiced when his new-found elixir restored his ruddy good looks. He was convinced he had discovered the Fountain of Youth, and increased his intake of Radithor to three bottles per day. Apparently treating his magical new potion more as a recreational drug (or even an aphrodisiac) than a medicine, he sent cases of the water to his friends, business partners, and even gave it to his racehorses. He was, however, exhibiting the first signs of radiation poisoning as his body over-produced red blood cells; the feelings of increased well-being dissipated, and he began to suffer headaches and loss of teeth; soon, his bones began to splinter and fracture. Medical experts from both sides of the 'mild radium therapy' dispute took up opposing views of Byers' mysterious condition. By the autumn of 1931, a Federal Trade Commission inquiry was underway into Bailey's promotion of his nostrum; there was no means by which Bailey could have been prevented from selling radioactive medicines, which was perfectly legal; the only option was to prove that his advertising claims were fraudulent. As Byers was far too ill to travel, an attorney was sent to his home on Long Island to record his evidence:

> A more gruesome experience in a more gorgeous setting would be hard to imagine. We went to Southampton where Byres had a magnificent home. There we discovered him in a condition which beggars description. Young in years and mentally alert, he could hardly speak. His head was swathed in bandages. He had undergone two successive jaw operations and his whole upper jaw, excepting two front teeth, and most of his lower jaw had been

> removed. All the remaining bone tissue of his body was slowly disintegrating, and holes were actually forming in his skull. [26]

The month after he gave his testimony, one of Byers' friends, to whom he had given Radithor as a gift, died from what was soon identified as radium poisoning. Bailey's defence to the Commission included the assertion that he was only emulating others in his promotion of radium-laced waters, pointing to the Hot Springs, Arkansas and Saratoga Springs, New York, both of which establishments were operated by government departments.

As we now know, and the public of 1930 was just beginning to recognise, radium is a 'bone-seeker'. While most ingested radium is excreted, some always migrates to the bones, where it remains continually emitting alpha particles, gradually destroying the skeletal structure, and causing destructive changes in the tissues and organs. That much had been understood by medicine two decades before Bailey began his activities; the trouble was that radium had become so much the plaything of unqualified swindlers like Bailey in an age when there was little, if any, regulatory framework to either constrain them or to record what they were doing.

Byers' body was disintegrating as a direct result of chronic radium poisoning. In January 1932, an order was issued against the continuing fraudulent claims of the Bailey Radium Laboratories, but it was too late for Eben Byers whose ravaged body was now far from maintaining the red-blooded pretence. The fifty-two-year-old suffered an agonising death from brain abscesses and acute anaemia in hospital in New York in March 1932. He had severe bone necrosis and his bone marrow and kidney functions had been destroyed. Despite the fact that he had consumed no Radithor for over a year, his breath, bones and internal organs were all highly radioactive; it was estimated that his body contained 360 times the amount of radium which would be allowed today for a registered radiation worker.[27]

Bailey was immediately sought out by press and public, and insisted that his product had nothing to do with Byers' death. He claimed, possibly quite correctly, that he had drunk more radium-water than any other man alive, and that it had done him no harm. The following day, when New York city officials tried to make contact with him, it was clear that he had fled the scene. Widespread removal began of all similar radium products from public sale, and the publicity began to unearth accounts of similar cases to that of Eben Byers. However, the shameless Bailey soon reappeared in association with several new radioactive products including the 'Adrenoray', a radioactive belt-clip; and the 'Thoronator', which he described as: 'a radioactive health spring for every home or office, as rich in vital rays as some of the most famous health springs of the world, and hundreds of times richer than the old-fashioned radium jars'. One of Bailey's collaborators was C. Everett Field (mentioned earlier as one of Joseph Flannery's first employees). In advertising and pamphlets, Field endorsed Bailey's activities in his capacity as 'director' of his own 'Radium Institute' in New York, and was quite open about his attitude to the promotion of radium treatments:

> From 1920 on to 1930 I had among my most considerate patients a most interesting group of wealthy patients from Maine to California.... I had Governors from three of our eastern states and a large group of army officers or their families.... Boy, it was fine sleddin' while it lasted. [28]

Bailey also sold the 'Bioray', a radioactive paperweight which he described as 'a miniature sun, far richer in the short, invisible rays than the sun':

> When the Bioray is placed in the room it immediately floods it with invisible gamma rays. Thus one can have these rays whenever one desires them, day or night, winter or summer, rain or shine.

> With Bioray on one's desk or at one's bedside one can obtain a steady flow of gamma rays continually, without any fuss or bother and without any interruption of the daily routine.[29]

This was a claim that conveniently ignored the fuss and bother suffered by Eben Byers.

Bailey went on to edit a small-town newspaper and ironically wrote a book on psychology; later, he was an aircraft observer during the Second World War and patented a number of military inventions. He died from bladder cancer in 1949 in Massachusetts at the age of sixty-four, having never been prosecuted for his commercial activities due to the lack of appropriate regulations and laws. To the end, Bailey insisted (from his wholly unqualified standpoint) that not only was radioactivity completely harmless, but that it was efficacious. And he used that favourite old metaphor:

> Radioactivity is one of the most remarkable agents in medical science. The discoveries relating to its action in the body have been so far-reaching that it is impossible to prophesy future development. It is perpetual sunshine.[30]

That might have been the end of the William John Aloysius Bailey story but for the fact that twenty years after his death his skeleton was exhumed for forensic examination under the auspices of the long-term and wide-ranging radiobiological studies being conducted by the Massachusetts Institute of Technology and the Argonne National Laboratory. His remains were still radioactive, and revealed significant change and damage as a result of his own activities.[31] Unwittingly, from the grave, the charlatan finally gave up some of the facts that have contributed to an understanding of the problems he helped cause. In particular, more information continues to be analysed on the radon problem (which is of concern in specific geological areas) and on the contentious proposition that long-term exposure to low-level radiation is acceptable or, according to some arguments, even desirable.

Although Byers' death was not the first attributed to radium, it was the first proven to have been caused to a member of the public by a readily available radioactive patent medicine used without professional supervision. While there was a strong public reaction, it was directed less against the relatively unknown mystery of radioactivity itself than to the more familiar and easily understood notion of dangerous quack medicine, with which history has been commonly and continually plagued. A much bigger, and much more far-reaching, scandal had its beginnings a very few years earlier in which radium played a central and deadly role. This would lead directly to the beginnings of the modern and socially vital science of occupational medicine, and to the concept of employers' liability for the protection of employees' health and safety at work, and for compensation of employees killed or injured while at work. The public enthusiasm for all things luminous and 'certified containing radium' was at the heart of the tragedy of the New Jersey dial painters.

NINE

UNDARK'S FATAL SHADOW

In the USA, between 1913 and 1915, an Austrian immigrant chemist and amateur artist named Sabin A. von Sochocky, who had trained briefly in Paris with the Curies in 1906, devised one of the earliest commercial luminous paints. He received $500,000-worth of financial backing from the Metals Thermite Co. and set up a small laboratory on 23rd Street in New York City.[1] His secret formula contained zinc sulphide, radium and the cheaper mesothorium; small amounts of other elements such as copper, manganese, lead, arsenic, thallium, uranium and selenium were also present. Von Sochocky was an enthusiastic promoter of his paints, foreseeing piano keys and the conductor's baton in a concert hall being coated with his luminous paint. He also predicted a time when rooms would be illuminated solely by the light thrown from walls and ceiling by luminous paint:

> Pictures painted with radium look like any other pictures in the daytime, but at night they illuminate themselves and create an interesting and weirdly artistic effect. This paint would be particularly adaptable for pictures of moonlight or winter scenes.

> The time will doubtless come when you will have in your own house a room lighted entirely by radium. The light thrown off by radium paint on walls and ceiling would in color and tone be like soft moonlight. [2]

Von Sochocky whimsically named his paint 'Undark' and with Dr George S. Willis established the Radium Luminous Material Co., which operated variously in New York at 535 Pearl Street, 55 Liberty Street, 15 Exchange Place, Jersey City, and in Delaware. By 1919, instead of radium (^{226}Ra), the company was using mesothorium (^{228}Ra), obtained in workable residues containing 50 per cent mesothorium and 50 per cent barium bromide, from two companies extracting thorium from monazite sand. In 1921, the company (by then renamed the United States Radium Corporation), acquired ore extraction rights in Paradox Valley, Colorado, and built an extraction plant and factory at 166 Alden Street, West Orange, in Essex County, New Jersey (close to Newark International Airport and not far from Thomas Edison's laboratory at Menlo Park in which some of the first X-ray injuries had occurred twenty years earlier). The companies employed girls and women to paint watch and clock dials, paying them \$20–\$25 a week at a time when they might have earned \$15 in an office.

Over a seven-year period the Radium Luminous Material Co. employed over 800 women as dial-painters. A huge range of spin-off products included luminous crucifixes, light-pulls, fishing lures and revolver sights, and the company's advertising carried the slogan 'I want that on mine'. For \$3 you could buy a do-it-yourself Radium Illuminating Set consisting of radium paint, adhesive, thinner, mixing cup and rod, and a camel-hair brush. The dial-painting studio was a well-organised operation where working conditions appeared good, with the dial-painters – almost all were women – seated at rows of desks in a large, well-lit, north-facing room on the second floor of the building. The research director of US Radium between 1921 and 1923 was

Dr Victor Hess, who became a Nobel Prize-winner in 1936 for work on cosmic rays; Hess retained his links with US Radium for many years after he left the company. Most of the employees were locals, often from extended families, and the working atmosphere was comfortable, friendly and well-paid. The fact that it was wartime, and the company was luminising watches for US troops, was probably an added incentive. The one difficulty the company initially faced was widespread complaints from nearby residents that heavy chemical fumes were being released from the radium refining plant. When the city threatened to close the premises, filtering equipment was installed in the extraction chimneys, and legal proceedings were dropped.[3] The plant processed about two tons of ore every day and large volumes of wastes and tailings were stored on the site before being transported and dumped on areas of undeveloped land in the neighbourhood.[4] Many of the dial-painters had come from the ceramics industry and those who had not were given appropriate training in the delicate hand-painting techniques. They were paid by results and were expected to complete over 250 dials in a day. They used stiff-haired brushes with only three or four hairs, the paint being contained in a small crucible holding about 10ml of the mixture.

> In painting the numerals on a fine watch, for example, an effort to duplicate the shaded script numeral of a professional penman was made. The 2, 3, 6 and 8 were hardest to make correctly, for the fine lines which contrast with the heavy strokes in these numerals were usually too broad, even with the use of the finest, clipped brushes. To rectify these broad parts, the brush was cleaned and then drawn along the line like an eraser to remove the excess paint. For wiping and tipping the brush the workers found that either a cloth or their fingers were too harsh, but by wiping the brush clean between their lips the proper erasing point could be obtained. This led to the so-called practice of 'tipping' or pointing the brush in the lips.[5]

The composition of the paints themselves affected the frequency with which they resorted to 'tipping' (which was a common practice in the china-painting industry). The commoner water-based version of the paint used a gum-arabic base which, although sweet-tasting, was sticky, causing brushes to clog and therefore to require more pointing with the lips; an oil-based version used glycerine and sugar, or amyl-acetate ('pear oil'), both of which were sweet, recognisable in taste and therefore possibly unobjectionable to swallow. Tipping with the lips was not a casual practice – it was taught by instructors (although the company later denied that it was a sanctioned technique). Every time they did it, the women were ingesting radium.

Sabin von Sochocky resigned from US Radium in 1921 at the time of the company's reorganisation, and started the General Radium Corporation, supplying radium for medicine; neither he nor his new company had any further business interest in luminous materials. For neither US Radium nor its employees was there any 'problem' in working at the radium dial-painting studio, and the women were given no particular instruction in safety precautions. Because of the well-known medical benefits that radium bestowed there was considerable attraction in working with the exciting new substance. The company even supplied a form of sand derived from the processing residues for re-use in children's play areas; it was surely positively healthy. One dial-painter did wonder why, after blowing her nose, her handkerchief glowed in the dark. But everyone knew it was all harmless and, of course, fun; during meal-breaks, the women would use up their left-over paint by painting their teeth, faces and fingernails and larking about in the dark. Often they would go home from work with their hair and clothes speckled and glowing in the dark.

It was authoritatively estimated that one worker painting 250 dials a day would have ingested 125mg of paint, assuming one 'tipping' per dial (and that was almost certainly an underestimate):

> If the worker pointed the brush a maximum of fourteen times per dial (and one girl stated that she often licked her brush twice for each watch figure) she would ingest about 1.75g per day. Therefore it is possible to state that a worker could actually ingest from 125mg to 1.75g daily, which would contain from 3 to 43 micrograms of radioactive substances. Working only six months a year, and only five days a week, from 360 micrograms to over 5mg of radioactive substances could be swallowed in that time. [6]

At the time, there was little understanding of the effects on the body of ingesting radium. The ability of the non-penetrating charged alpha particles to irradiate the body's internal organs – in particular, bone – was unrecognised, as was the potential for consequential biological changes. It was widely assumed that the body efficiently excreted such material.

Assisted by US Radium, which began to concentrate mainly on producing the paint, many clock companies set up their own dial-painting studios up to the late 1940s. In Connecticut, there were companies at Bristol, New Haven, Thomaston and Waterbury; there were others at Bloomsburg, Pennsylvania; Newark, New Jersey; and at Athens, Georgia and elsewhere. The biggest competitor to US Radium was the Radium Dial Co. at Ottawa, Illinois, about 80 miles south-west of Chicago. The company originated in 1917 in Long Island City, and opened studios in Peru and Streator, Illinois, before settling in Ottawa in 1923. The business had been founded by Joseph Flannery, founder of Standard Chemical and its various subsidiaries; he had died after years of enthusiastically promoting the use of radium – including digging it into his tomato garden and consuming the produce. His son-in-law, Joseph Kelly, took over Flannery's various radium businesses, and set up Radium Dial in a former high-school building. Here he employed several hundred girls, mostly painting dials under contract for the giant Western Clock Co. (more popularly known by their trade-mark name

'Westclox'). As at West Orange, the employees felt they were in a comfortable, secure and happy working environment.

By the end of the First World War, the US was producing over 4 million watches a year with radium-luminous dials, but production of military materials quickly gave way to civilian 'novelties'. In the mid-1920s there were about 120 luminising plants across the USA, employing more than 2,000 dial-painters, and the techniques, including lip-pointing, devised by US Radium (as the market leader) were universal. All went well – for a while. But soon the time of innocence came to an end. Between 1922 and 1924, nine young West Orange women dial-painters died, and a further twelve suffered unexplained and devastating illnesses; every year the numbers increased. Death certificates recorded various causes including anaemia, syphilis, stomach ulcers and necrosis of the jaw. However, no autopsies were performed, and therefore no common factors were authoritatively identified or suspicions aroused. Something was going badly wrong.

In early 1924, the Consumers' League of New Jersey, a voluntary community organisation concerned with the employment of women and children, examined working practices at the West Orange factory. They found conditions to be generally good, although they questioned fumes coming from the plant. The US Public Health Service and the New Jersey State Department of Labor were consulted, but took no action.[7] Helen Wiley, the perceptive Consumers' League secretary, noted that four of the dead women, and some of those who were ill, had undergone operations on their jaws, and enlisted the help of the National Consumers' League. Support came from Alice Hamilton, a doctor dedicated to getting industrial health reform on the public agenda. Hamilton had just achieved success in a campaign to protect workers at General Motors against the neurological damage caused by the use of the organic compound tetraethyl lead in gasoline. She had the help of Walter Lippman, Editor of the *New York World* – a respected, campaigning liberal newspaper founded by Joseph Pulitzer. To both of them, the case of the dial-painters was: 'one of the most damnable travesties of justice that has ever come to our attention'.[8] There was to be no easy public victory, however; layers and layers of intertwining politics – civic, state, federal, legal, medical and commercial – were destined to resist the changes necessary to protect or compensate the workers.

The breakthrough began in the autumn of 1924. Theodore Blum, a New York dentist, recognised similar severe jaw problems in a number of his patients who worked at the West Orange factory, and surmised that occupational poisoning may have been the cause. Other dentists had encountered similar cases and attributed them to 'phossy jaw' – a problem recognised in workers who handled phosphorus. Blum read a paper before the American Dental Association, which published his findings in its journal. This account came to the attention of Dr Harrison Martland, pathologist and Chief Medical Examiner for Essex County, whose role in the early diagnosis of radium poisoning was to be pre-eminent. He knew Blum by reputation and was impressed by his proposition that there was a new and serious form of occupational health hazard at work.

In the summer of 1925, Martland conducted an autopsy on Dr Edward Lehman, the US Radium Corporation's chief chemist, who had died suddenly. There was extensive radiation damage, particularly to the bone marrow, liver and lungs. His bones were so radioactive that, left on a photographic plate in the dark, they photographed themselves. Lehman had not swallowed any paint; he had merely breathed the air in the factory. Uniquely, the refining plant and painting studio were situated close together, in the same premises, and this was thought to have exacerbated any problem. The company suppressed the autopsy report but shortly afterwards, Martland, working with the unconditional assistance of Sabin von Sochocky, had the first opportunity to perform a post-mortem examination on a US Radium dial-painter. The woman survived in hospital for only a week before dying in

July 1925: she had severe anaemia and her mouth and gums bled spontaneously. Her body was radioactive and her breath, when feebly exhaled, made a zinc sulphide screen glow. Essentially, Martland found high levels of radioactivity in her bones, spleen, liver and – most crucially – bone marrow, the tissue in which blood cells are produced.[9] Martland was able conclusively to report the detection of gamma rays in living workers, and the exhalation of radon from their lungs.

Dr Frederick Hoffman, a statistician with the Prudential Life Insurance Co., was asked by the National Consumers' League to make an assessment of the US Radium operation. Sabin von Sochocky, still concerned by the steady drip of allegation and insinuation about the continuing US Radium activities, was distressed by the problems affecting his former employees. He resolved to help Hoffman, whose report published in the *Journal of the American Medical Association* in September 1925 first focused attention on the 'tipping' of brushes, and concluded that the deaths and illness were unlikely to be coincidental. Hoffman was sure that deaths would continue unless practices were changed and insisted that medical claims, being clearly occupational in character, should be brought under legislation on compensation for industrial diseases. He made appreciative reference to von Sochocky, whose 'thoughtful but highly technical observations make a valuable contribution to the strictly scientific study of the subject.' Hoffman was entirely convinced that:

> the situation is one of extraordinary complexity. Apparently, radium necrosis occurs only under certain and quite exceptional conditions. It is not the fact of general exposure to radioactive substances or nearness thereto, but, apparently, the direct result of introducing such substances in minute quantities into the mouth through the insanitary habit of pencilling the point of the brush with the lips.
>
> Every case personally investigated gives an unmistakable history of this habit, while the numerous roentgenograms clearly

> indicate the consequences to both the roots of the teeth and the jawbone. [10]

Individual stories are poignant. One very ill New Jersey woman, Edna Hussman, waking in the middle of the night to take her medicine, caught sight of her reflection in a mirror and collapsed in a dead faint. What she had seen was a luminous glow radiating from her body. Illness and death also visited Ottawa with suspicious frequency. Margaret Looney had a tooth extracted and the wound not only failed to heal but revealed severe honeycomb destruction of her mouth and jaw. Her parents suspected her illness was the result of working at Radium Dial and took her to a specialist in Chicago, who reluctantly confirmed their suspicions, but nevertheless told them, 'I cannot speak out and tell you because this would be the end of my career.' [11] Acting in some secrecy, the US Radium Corporation arranged for experts from the Harvard School of Public Health to report on conditions at their factory. Even with their poor understanding of radiation, they recognised a mess. Not only walls, floors and work surfaces, but workers and their clothes were flecked with radium paint. Dust samples glowed in the dark and gamma emission was five times greater than expected. Not a single worker tested had a normal blood count. The Harvard report concluded that workers were exposed to excessive radiation, from both external and ingested sources. The company – which had asserted that 'radium in small doses is a stimulant' – suppressed the report, but quietly told supervisors to prevent the licking of brushes.[12] Far from publishing the critical report, Joseph Kelly issued a statement in the press on 7 June 1928:

> We have at frequent intervals had thorough physical and mental examinations made by well-known physicians and technical experts familiar with the conditions and symptoms of the so-called radium poisoning. Nothing ever approaching such

> symptoms and conditions has ever been found by these men – on the contrary, they have commented on the high standard of health and physical appearance of our employees and the excellent conditions in which they work. If the report had been unfavourable, or if we at any time had reason to believe that the work endangered the health of our employees, we would at once have suspended operations. The health of the employees of the Radium Dial Co. is always foremost in the minds of its officials. [13]

When Margaret Looney eventually died after two weeks in a company hospital, having been allowed no visitors, the civic and medical authorities demanded instant burial in the middle of the night. The family refused and later, when their pathologist arrived to conduct an autopsy, he discovered that one had already been done in secret by the company, and was told that she had died from diphtheria. Years later, when the body was exhumed for examination at Argonne National Laboratory, it was found to have been encased in lead, with a transparent window for visual inspection – somewhat unusual for a diphtheria victim.

In 1928, the president of US Radium wrote a long letter to the New York Commissioner of Public Health, in which he blatantly sought to counteract some of the terrible publicity that his company was attracting:

> The [dial painting] work was easy, the operators well paid, and as conditions turned out, we unfortunately gave work to a great many people who were physically unfit to procure employment in other lines of industry. Cripples and persons similarly incapacitated were engaged. What was then considered an act of kindness on our part has since been turned against us, as all previous employees, regardless of what they may have been suffering from or are suffering from at the present time can be attributed to 'Radium Poisoning'. [14]

Von Sochocky, who by this time was himself beginning to suffer ill-effects, issued a statement advising all those who had worked with his luminous materials immediately to undergo a thorough physical examination. The man who had so lightheartedly promoted the benefits of his luminous paint now adopted a highly responsible and selfless response to what was clearly a hideous problem. He had himself been heavily exposed during his supervision of the refining process. On at least four occasions he had been present when tubes containing radium or mesothorium of high purity had exploded. He appeared to have been enthralled by the substance, according to a report for the US Department of Labor:

> He was fascinated by the qualities of radium, and is said to have played with it, taking the tubes of radium salts out of the safe and holding them in his bare hands while watching the scintillations in the dark. He is also said to have immersed his arm up to the elbow in solutions of radium or mesothorium. [15]

Martland began to realise that radioactive material was not rapidly excreted from the body, as had been thought. It was now clear that such material accumulated in certain organs and continued to irradiate surrounding tissue. Specifically, radium was identified as a 'bone-seeker'. Research everywhere was widened to include people who had been given a variety of medical and non-medical treatments involving radium and radon. There was no shortage of potential subjects, given the numbers of people who had used the various radium-water preparations.

Martland, who conducted considerable research into existing recorded medical cases arising from the handling of radioactive substances, warned the medical profession; the implications not only concerned those with an occupational hazard, but applied to hundreds of thousands of doctors and patients who were using radium for what they thought were efficacious purposes. Despite these findings, the US Radium Corporation continued to deny any culpability, blaming the deaths and illness on poor personal hygiene. It was denied that radium could be the cause, since dial-painters in Europe had not suffered in the same way. What was ignored was that in Switzerland and France brushes were banned, the paint being applied with the use of fine glass rods. Having been disappointed by Hoffman's concerns, the Radium Corporation hired a new consultant, Dr Frederick Flinn, Professor of Industrial Hygiene at Columbia University. He had a clear understanding of the potential volumes of radium ingested during mouth 'tipping' after comparing procedures with one factory where brushes were pointed by pulling through cloths, enabling accurate measurement of deposits. Despite this knowledge, Flinn initially obliged the company by offering anodyne assurances that there was no hazard to the dial-painters. He disputed theories that radioactive particles could have become lodged in the mouth and therefore caused necrosis, 'because my inquiries indicate a general use of the tooth brush among the employees'.[16] He insisted that any small amounts of radium ingested were assuredly excreted within a few days, and that the deaths had been due to 'bacterial infection'. Only later did he reluctantly acknowledge the possibility that 'radioactive material is at the bottom of the trouble'.

Sabin von Sochocky suffered severe destruction of his jaw, mouth and hands. Since 1925 he had given Martland and Hoffman enormous help in examining his workers. At one point, he exhaled his own breath over a screen of phosphorescent zinc sulphide and produced a greater radioactive effect than any of the dial-painters on whom they were working. His fate was horribly clear. As Martland recorded:

> From 1913 to 1921 he had been exposed to intense radiation from radioactive substances, during which time he had personally extracted 30g of radium from the ore, being exposed continually to heavy penetrative radiation. He was exposed to the inhalation of dust in the crystallising laboratories. From 1919 to 1920 he was

> exposed to heavy radiation in a small room from large amounts of emanation by inhalation and ingestion and to highly concentrated dust from radium, mesothorium, and the like. On four occasions he was exposed to explosions of tubes containing high concentrations of radium and mesothorium. Since 1921 he had occasionally been exposed to external penetrative radiation from radium used for therapeutic purposes.[17]

Von Sochocky's bone marrow and red blood cells were being steadily destroyed, although he was able to continue a more or less normal life until August 1928, when he became extremely weak and anaemic. Despite a long series of blood transfusions, and the consumption of considerable amounts of raw liver in an attempt to improve his blood condition, he died of aplastic anaemia in November at the age of forty-six. Von Sochocky may have been one of the enthusiastic radium promoters, but when he understood the implications, he behaved with an utterly self-sacrificing devotion to helping others. A vast archive of the papers of von Sochocky and US Radium was made available to the authorities as part of what was to become an extended inquiry. Although much of this documentary material was declared radioactive and destroyed, it was all microfilmed for archival purposes. In his own detailed records, Martland gave Sabin von Sochocky moving recognition:

> He died a horrible death, similar to that seen in chronic benzene poisoning, the acute leukaemias and other acute blood dyscrasias. During the time I knew him he gave all that was in him to help and comfort others suffering from this disease. Without his valuable aid and suggestions we would have been greatly handicapped in our investigation.[18]

In 1929, the US Department of Labor produced a substantial report on the dial-painting industry after the inspection of thirty-one plants; while recommending guidelines for workers' protection, it dramatically recommended that radium dials were unnecessary and that their manufacture and import should be prohibited. The same year, the US Public Health Service was asked to begin a national survey of the watch and clock industry to determine the full extent of the occupational risk, but as Martland was not slow to point out:

> Such an investigation would be of great importance, no doubt, and it is possible that similar disastrous effects may be revealed. The great trouble with most investigations is that they always start after the harm has been done.[19]

Apart from confirming that radium was not necessarily excreted from the body, Martland was the first doctor to recognise two factors in relation to radiation damage, both of which have remained characteristics of the problem. The first was the accumulation effect and the second was that diagnosis was often impossible and inconclusive until years after initial exposure:

> The late development of symptoms, often occurring from one to seven years after the patients leave the employment, and their resemblance to various other diseases would make a diagnosis almost impossible if the patients had moved to another locality. Physicians not aware of this condition might treat a victim for sepsis, anaemia, Vincent's angina, rheumatism or 'God knows what'. [20]

The dial-painters' disastrous predicament quickly became the subject of intense and sympathetic public interest. Some of the headlines were lurid (the women being referred to as the 'Living Death Victims', or the 'Legion of the Doomed'), and became increasingly so as court actions were opened against the US Radium Corporation and the Radium Dial Co.. The widow of Dr Lehman began legal action, as did the relatives of many of the women who had died. Several of the dead were exhumed in order that relatives might have the bitter satisfaction of eventually knowing that death had been due to radiation poisoning rather than 'God knows what'. But as Martland's detailed pathological evidence mounted, groups of workers who were effectively under sentence of a highly unpleasant death, grouped together to fight for damages – which were not available under existing compensation laws. In the main group action against US Radium, each woman claimed $250,000. The newspapers had a field day. 'THE FIVE WOMEN DOOMED TO DIE' became celebrities in what was a long and dirty fight for fair treatment. Some of these women had undergone distressing surgery and were suffering the most painful and crippling conditions. Several had to be carried into court; one could not even raise her hand to take the oath. Harrison Martland – angered and frustrated by the frenzied publicity and misrepresentation of his views – has been unfairly criticised for abandoning the dial-painters. He maintained that, although US Radium had deplorably and continually refused to admit that there was any poisoning taking place and should not therefore assume any responsibility, they had acted honestly and without any criminal intention; he therefore resisted pressures on him to assist in the prosecution of the company. Although he no longer acted in the direct interest of the dial-painters, he continued to study their case-histories and to collaborate with their physicians and others in identifying medical outcomes and in promoting the introduction of proper workers' compensation laws.[21]

In the action against US Radium in New Jersey, the court convened around the deathbed of one woman. The front pages of the newspapers printed large photographs with headlines such as, 'LIVING DEATH QUIZ AT BEDSIDE'. Although interest in the case achieved international recognition, it appeared likely that legal delays would extend indefinitely. The company decided not to fight on the substantive issue, but tried to halt proceedings on the grounds that the state's Statute of Limitations required that claims had to be filed no later than two years from inception of the disease. This provoked uproar; public sympathy soared, quacks and faith healers joined the fray, as did candidates in the 1928 presidential election campaign. In the end, a federal judge unconnected with the case stepped in as an intermediary. Within days, he had brokered an out-of-court settlement; each woman received $10,000 as a lump sum, plus an annual pension of $600, together with legal and life-long medical expenses. The New Jersey plant closed shortly afterwards, although the company continued to operate elsewhere.

Similar settlements were reached with Joseph Kelly's Radium Dial Co. in Ottawa, Illinois. Although the company closed its plant, it opened a new factory six weeks later only two blocks away. The Luminous Process Co. assured its workers that there were no problems in working with radium. What it didn't tell them was that its owner was Joseph Kelly. That plant continued in operation until 1978, when it was closed by regulatory authorities for continual breach of health regulations. At that time it was estimated that there were over eight million original radium watches and clocks still in circulation, delivering a collective annual radiation dose greater than that from all US uranium mills, reprocessing plants and nuclear facilities. It was as late as 1963 that authorities in New York banned the sale of pocket-watches with radium dials, which were described as having 150 times the permissible radioactivity. The rationale was that, since dose to any tissue varied inversely with the square of distance, pocket-watches should be banned because they were usually worn close to the waist, with faces towards the body, and therefore posed a genetic risk to the sexual organs; wrist watches 'generally faced away from the body'.[22]

Harrison Martland felt that legal protection for workers was more important than condemning the employers, whose real offence was ignorance (although he deplored their blind refusal to accept what had actually happened). After about 1927, there were few workers who acquired classic radium-induced tumours, and it is probable that radium was the first carcinogen to attract quantitative protective standards. Nevertheless, Martland raged at the law's delays and the fact that people with no legal or occupational protection had spent their life savings in futile attempts to protect themselves:

> All these cases should properly have come under the compensation law. This would have saved the company an enormous amount of worry, loss of business due to publicity, and the possibility of impending financial ruin. If this disease had been properly covered by law, the victims would have been saved from the ignominy of becoming objects of charity. A competent body of medical authorities after the evidence was submitted could have disposed of the case in a day's time without the utter ridiculousness of dragging a simple case along for years in the courts. The only advantage out of this terrible mess has been, as usual, to the legal profession. [23]

Plus ça change.★3 (In matters of occupational health, the fractured relationship between workers and employers, and the ability of lawyers to be the principal beneficiaries, still continues; e.g. disputes concerning benzene, asbestos, organo-phosphates and, of course, radiation.)

In later years, from 1924 to 1954, the US Radium dial-painting plant was on five floors of a ten-storey building on Pearl Street in lower Manhattan, while the main laboratory was in a penthouse in Lafayette Street, between City Hall and Chinatown; here, a range of radium products and neutron sources was made, and radium reclamation was undertaken. One new product advertised in 1947 was a semi-transparent, moisture and acid-resistant quarter-inch diameter luminous plastic tubing.[24] This had originally been produced for military use, as a marker for night practice parachute drops and ships' passageways. After 1954, the company's operations were briefly transferred to Morristown, New Jersey, and Bloomsburg, Pennsylvania.[25]

After the company diversified into the production of a wide range of luminous and medical products, it developed a number of subsidiaries, including the Luminite Corporation (New York City), the Radium Aluminium Manufacturing Co. (Newark, New Jersey), the Radium Application Co. (Newark), and two watch companies the Keystone Watch Co. (Newark and Jersey City) and the Capital Crescent Watch Co. (Newark).[26] The plant at Orange, New Jersey, was sold in 1943 and processing was maintained in Bloomsburg (Pennsylvania), Bernardsville and Whippany (New Jersey), and North Hollywood (California). In 1956, an attempt was made to re-enter the radium market by establishing a Canadian venture, Ratalin Kirk Ltd, in Toronto, but this move was abandoned within two years; there was also a little-known subsidiary in Switzerland.[27] Some of the many parts of US Radium collapsed after Belgian and Canadian monopolies came into being, but the company remained alive (manifesting numerous name changes) in Bloomsburg. The US Radium name finally disappeared in 1980 when its plant in Bloomsburg was renamed the Safety Light Corporation, luminising with tritium and allegedly investigating new uses for uranium wastes. The name may have changed, but radioactive wastes were dumped in silos and lagoons, polluting the nearby Susquehanna River.[28]

In Britain, much less luminising work was carried out than in America; even so, there was a better understanding of potential danger. The British government officially declared radium to be a hazardous material in 1916, and published warnings from 1921. Sticks rather than brushes were used, thereby avoiding the 'tipping' problem (brushes were prohibited by statutory order from 1941). The paint used in Britain was based on the foul-tasting solvent toluene, rather than the accidentally attractive and sweet-tasting gum arabic used in the USA, and some plants also used gloves or glove-boxes. There were a few small plants during the First World War, with a peak of production during the Second World War totalling between thirty and forty small luminising plants, mostly associated with car factories or military depots. Unlike America, there were no early studies or records of the dial-painting industry before the Second World War. As the war began and it was realised that the industry would be greatly expanded, the Ministry of Labour established regulations and supervision over the industry in Britain, and regular clinical and blood tests became mandatory. After the war, production involved the less dangerous mesothorium rather than radium, and other safer radionuclides were soon introduced; nevertheless cruder radium compounds were in use in until the late 1960s and are still in storage today.

Army and Ministry of Labour records supplied the basis for studies beginning in the late 1950s of British workers and luminising methods. Records of about 1,900 workers (mostly women) were researched and about 1,600 workers were traced. It was generally found that there had been no serious radiation damage comparable to that in America, although results in 1981 showed that 'those who were under thirty years of age when they started work show a significantly increased risk of dying from breast cancer.' In all, 78 per cent of the women were in that under-thirty category. Inevitably, there were slack practices, much less knowledge, and fewer protection measures than apply today:

> The attitude towards radiation protection in a luminising workshop is nearly always far more relaxed than that found in a modern, well-run, radiochemistry laboratory.[29]

Long-term research into the plight of the US dial-painters had begun in the 1920s at Harvard and, notably, at the Massachusetts Institute of Technology under Professor Robley Evans. In 1946, studies that had been undertaken at the Metallurgical Laboratory of the University of Chicago were transferred to a new federal nuclear research laboratory (operated by the university on behalf of the government) at Argonne, 75 miles from Ottawa, Illinois. Over a period of many years a lot of the surviving dial-painters were the focus of research to quantify the effects of radiation on the human body, and the corpses of some of those dial-painters who had died were exhumed as part of the studies. It has been incorrectly claimed that little effort was made to explain very much to the women; while it is true that there was no question of medical treatment being offered at Argonne, all of them were given access to as much information as they wished, and detailed reports were regularly given to their own doctors. In all, 2,400 individuals had their body radium content measured, many of them over most of their adult lives until death, in the largest ever study of the effects on humans of an internally deposited radioelement. Despite the extremely detailed investigations and analyses, the conclusions have left questions unanswered, and have given rise to apparent anomalies:

> No symptoms from internal radium have been recognised at levels lower than those associated with radium-induced malignancy. Radium levels 1,000 times the natural ^{226}Ra levels found in all individuals apparently do little or no recognizable damage. These statements may suggest that a threshold exists for radium-induced malignancies; at least, they recognize that the available data demonstrate a steep dose response, with the risk dropping very rapidly for lower radium doses. [30]

As the years passed, there was some professional interest in following up the incidence of the birth defects and long-term genetic change and damage which appeared in the families of some of the dial-painters, but very little work was officially sanctioned. Somewhat controversially, the US Department of Energy discontinued the Argonne radium study in 1993, despite the fact that there were more than 1,000 individuals still alive with recorded radium burdens who had been expected to be subjected to continuing assessment until death.

The studies by Evans and his colleagues of the radium-induced tumours found in the dial-painters led directly to the establishment of a long succession of radiation protection standards. These benefits were apparent before the Second World War (when radium dial-painting increased) and continued as new elements came to be assessed. Plutonium, in the form of the fissile isotope plutonium-239 produced from uranium-238 in nuclear reactors, had huge implications for weapons production and power generation, but there was considerable anxiety about the potential for workers handling plutonium to acquire alpha-emitting radioelements in their bodies. (Plutonium was first identified in Chicago in 1940 by Glenn Seaborg who, thinking that it would be the heaviest known element, initially considered such possible names as Extremium and Ultimium before adopting the planetary precedents and honouring the 1930 discovery of the planet Pluto.[31] While Seaborg and others acquired plutonium by nuclear bombardment in a reactor, this most deadly radiological poison exists in nature in pitchblende, occurring in amounts equal to five parts per million million.) It is largely due to Robley Evans' earlier studies of radium victims, and the resultant controls, that there was relatively little cancer among plutonium workers during and after the Second World War, despite the fact that it is far more dangerous than radium; this was characterised to me by a radiation biologist as 'the epidemic that didn't happen'.

The project to assess radiological data gathered momentum in the post-Hiroshima period, when there were considerable efforts to incorporate 'field information' from Japan into protection standards. During the Cold War, there were undisclosed instances when scientific arrogance and political zeal apparently ran out of control. One experiment at the end of the Second World War at the University of Chicago involved a dozen women being injected with radioactive thorium dioxide (Thorotrast, used at the time as an X-ray contrast medium) just before giving birth. There was at least one early death, and half of the children still had measurable radioactivity in their bodies twenty-five years later. Demands for scientific investigation were routinely rebuffed, and little information has ever reached public notice. One informant observed to me, 'Medical arrogance doesn't need politics; politics just gives it more resources.'

While there had been publicity about similar events for a number of years, public anxiety was raised in 1987 after Eileen Welsome, a journalist working for a small-town newspaper in New Mexico, uncovered a footnote in records from a USAF weapons laboratory about eighteen people secretly injected with plutonium without their consent. She continued to uncover records that showed that thousands of prisoners, mentally retarded children, terminally ill patients and uranium miners had been subjected to similar experiments. These records had previously been unearthed by Dorothy Ligoretta, who had worked in the laboratory of Joseph Hamilton, a radiation biologist with a zeal for experimentation who worked on the Manhattan Project to build the first atomic weapons. The US Atomic Energy Act of 1946 had ensured the classification of an estimated 32 million documents, many of which have been found to contain evidence of an horrific and systematic abuse of American citizens, without either their knowledge or consent. The most they had been told was that they were involved in dietary research programmes. Different groups had been given food or drinks containing radioactive materials; some had been subjected to excessive X-rays; others to the injection of plutonium; others again had their reproductive organs exposed to various radioactive sources. Plutonium was so secret that the preparation used for injection was known only as 'the product'. Some of the estimated 200 separate experiments were carried out inside secret government nuclear establishments across the USA, but others were conducted by medical researchers in thirty-three civilian hospitals throughout the country.

Experiments conducted by Dr Eugene Saenger in Cincinnati involving intense 'whole body' radiation, allegedly given to often poorer, African-American cancer patients as therapy, were in fact conducted for military purposes. The eighty-eight patients were denied dedicated medical treatment and an unknown number died as a direct result. In 'Project Sunshine' (that homely, sunny metaphor again) Dr Willard Libby of the University of Chicago canvassed the acquisition of the corpses of six thousand babies (preferably stillborn or newborn) from around the world, including Britain, for strontium-90 experiments. Libby was quoted as saying, 'if anybody knows how to do a good job of body-snatching, they will really be serving his country.'[32]

Following the fire at Britain's Windscale nuclear reprocessing plant in 1957 (cynically renamed Sellafield after the incident), when radioactive iodine was released to the atmosphere in large amounts, experiments in Waltham, Massachusetts, recruited children for injection with radioactive iodine and calcium contained in safe-sounding Quaker Oats by offering them membership of a 'Science Club'; all records have been destroyed and no follow-up studies were conducted. It is claimed that unsuspecting populations in Nevada and Utah were deliberately exposed to radioactive fallout released during bomb-tests and by deliberate emissions from military nuclear plants; some of these activities almost certainly involved the use of 'volunteers' who were placed in potentially lethal locations.[33] It is unlikely that these events constituted legitimate 'experiment' since, so far as is known, there was no prior strategy and no eventual assessment.

Joseph Hamilton, although involved at the University of California in radiation experiments on humans and in the use of radiation as a weapon of war, wrote a memorandum on 28 November 1950 to the director of biology and medicine of the Atomic Energy Commission. He had begun to have second thoughts, and warned that such experimentation on uninformed humans was probably unethical, illegal and a breach of the Nuremberg Code of 1947:

> I feel that those concerned in the Atomic Energy Commission would be subject to considerable criticism, as admittedly this would have a little of the Buchenwald touch. The volunteers should be on a freer basis than inmates of a prison. [34]

The Nuremberg Code on medical experimentation on humans demands notification and follow-up and specifically insists that: 'the duty and responsibility for ascertaining the quality of consent rests upon each individual who initiates, directs, or engages in the experiment'. Hamilton's memo had no effect and the experiments continued at least into the 1970s when, after some scientists at Argonne wanted to ingest a harmless plutonium isotope and donate their bodies to scientific *post-mortem* examination, the authorities summarily rejected the idea. Dorothy Ligoretta was killed in a suspicious car accident when about to publish her account of these experiments. Later, Eileen Welsome won the Pulitzer Prize for her substantial book detailing her researches into experimentation on humans. When Hazel O'Leary became US Secretary of Energy in 1993, she was shocked to be briefed on the huge scale of such activities and referred to Hamilton's 'Buchenwald' comment when announcing a wholesale public disclosure of the 840,000 pages of relevant records, which are now available on the website of the US Department of Energy.

Similar activities were conducted in Britain, but lacking adequate freedom of information in this country, details are scarce. CND in Britain reported in 1996 that, while radiation experiments on humans began officially in the early 1960s, they probably began in 1957 and continued until the late 1980s. These activities were carried out at Aldermaston and Harwell, often in association with the US government's programmes. Ironically, it is the US Freedom of Information legislation that has revealed many of the available British details, proving facts that are at variance with statements made by the UK government. It seems clear that radioactive substances have been inhaled, injected, eaten or otherwise ingested by human subjects; radioactive isotopes used allegedly include those of niobium, barium, palladium, chromium, strontium, iodine, plutonium, americium, calcium and tritium. Such experiments would have circumvented internationally agreed guidelines on the use of radioactive substances in medical research; they were apparently sustained by a system for denial of official responsibility if anything went wrong. Successive British governments have never informed Parliament of such continuing experiments. Channel 4 television reported in 1995 that from 1955 to 1970, in the course of six thousand observations to determine radiation uptake following nuclear tests, the bones of dead children up to the age of five were systematically removed (femurs were preferred) without parental knowledge or consent. In Coventry, non-English-speaking Punjabi women were unwittingly fed radioactive iodine in 'special' chapatti flour provided by the Medical Research Council as part of a nutrition experiment; the chapattis were delivered to the door each morning and someone always turned up later in the day to ensure that they had been eaten. When the women were subsequently examined, the 'hospital' they were taken to was the nuclear laboratory at Harwell. Dr Alice Stewart, the renowned radiation epidemiologist, observed that the idea of this procedure constituting a medical treatment was 'a plain lie'.[35] In Oxford, whole-body radiation was administered in experiments similar to those in Cincinnati; Dr Alice Stewart suggested that giving such high intensity radiation to 'kill cancer' was not therapy, since it would undoubtedly also kill the patient.[36] Pregnant women in Aberdeen were repeatedly injected with radioactive iodine, in exchange for better food and private hospital rooms with television sets, but without the benefit of truthful explanation; in Liverpool and London, it was radioactive sodium. Inevitably, experiments that were conducted at Aldermaston – a military facility – have remained even more completely secret.[37]

It is clear that the problems brought about by the interaction of radiation and humans did not go away even after Martland and the doctors who followed him began to learn how to protect people. When military interests and political secrecy took over, as we have seen, medical integrity was often compromised. This was worse than medical paternalism; it became a pernicious form of bureaucratic fascism. (Such dishonesty has always existed and still continues; in March 2004 it was reported that bodies donated to New Orleans University Medical School had been sold to the US Army to be blown up in experiments into designing protective footwear against land mines.[38])

In Britain alone, there are probably hundreds of radium-contaminated sites which still pose some danger to the public. In earlier decades, there were no planning controls, safety procedures or occupational protection, and no formal record-keeping. By definition, the pioneering days of the production and use of radium used methods which were, by today's standards, unsophisticated in the extreme. Every activity involving man-made radiation has produced a disposal problem which is arguably no nearer a solution today than ever. Despite the many benefits which radiation technology has brought – particularly in medicine – earlier ignorance and more recent concealment have combined to produce mistrust and controversy. Chapter 11 will detail some of these issues, but first we have to follow the trail of what happened to the radium story after the period of the dial-painters and the difficulties of finding and exploiting the necessary mineral resources.

TEN

THE WORLD CHANGED FOREVER

In the aftermath of the dial-painters' scandal in the USA, which rumbled on through the course of a variety of legal processes in the late 1920s, the use of radium in luminous materials for the general public began to wane, as did the promotion of the various radium-waters and other non-medical products. After examinations conducted by the Bureau of Chemistry, the US government began to issue warnings in 1926 about the supposed medicinal efficacy of many items on sale:

> The products analysed by the Bureau included hair tonics, bath compounds, suppositories, tissue creams, tonic tablets, face powders, ointments, mouth washes, demulcents, opiates, ophthalmic solutions, healing pads and other preparations in solid, semi-solid and liquid form. Only 5 percent of the products analysed contained radium in sufficient quantities to entitle them to consideration as therapeutic agents and then only under very limited conditions. Highly exaggerated therapeutic claims are being made for many of these products.[1]

One of the radioactive devices that had particularly infuriated the authorities was a short glass rod coated at one end with a yellow substance and enclosed in a glass bulb. The gadget was supposed to be hung on the end of the user's bed where, the manufacturers claimed, 'it disperses all thoughts and worry about work and troubles and brings contentment, satisfaction and body comfort that soon results in peaceful, restful sleep'. [2]

By the late 1920s, radium processing for non-medical use was becoming a rarity in Britain, and in the USA was still very much in the shadow of the exploitation of the Congolese resources. In Canonsburg, Pennsylvania, the last remaining chemist working for Standard Chemical set up the Vitro Manufacturing Co. to process carnotite ore for radium, uranium and vanadium. Much of the processing was done using the abandoned tailings left in the town by Standard Chemical; some of this material was actually dug up from the town's road-beds, where it had been rolled down as discarded waste. Most of the output was used in the manufacture of coloured glazes for the ceramics industry, although some radium was produced until about 1942, by which time the main effort was devoted to vanadium extraction for use in steel production.[3] In the USA, there are recognised periods for the usage given to radioactive minerals: 1871–1905 was glass and ceramics; 1905–25 was radium; and 1925–45 was the vanadium period. (Two uranium periods would follow, an essentially secret period of government activity from the end of the Second World War to about 1967, followed by a more openly commercial period.)

The various charitable institutes and other bodies in Europe and the USA were still in business between the wars, acquiring radium supplies for medicine and surgery as best they could (mostly from Belgium). In London, the Radium Institute was still flourishing in 1925:

> [It] is probably the best equipped institution of the kind in the world. It possesses 6g of radium bromide, the money value of which is said to be almost $500,000. Since the opening of the

> Institute, about 10,200 patients have been treated, and 4,500 have been treated in hospitals with emanation apparatus. [4]

In 1929, there was talk in the House of Commons in London of the British government sponsoring a new company to attempt to exploit Portuguese ore resources, but bigger international events were to overtake such considerations. The Radium Sub-committee of the Committee of Civil Research examined the whole radium situation and decided that a new Radium Commission should be established to standardise treatment and enable security of supply and distribution.[5] There was still constant anxiety over the price and sources of supply of radium. The following year, the newly formed Radium Commission took its first steps to consolidate and regulate the use and supply of radium throughout the UK, thus avoiding duplication of effort and destructive competition between hospitals which ought instead to be collaborating. New post-graduate teaching centres were set up at the London Radium Institute and at Mount Vernon Hospital in Northwood:

> In accordance with the policy of concentrating national radium at a limited number of centres with medical schools attached, twelve 'national radium centres' have been nominated. Of these, Birmingham, Cardiff, Edinburgh and Aberdeen have already been supplied with radium; Manchester, Glasgow and Newcastle will receive a supply almost immediately; while agreements with Bristol, Dundee, Leeds, Liverpool and Sheffield have yet to be signed. [6]

As part of the centralisation process, national radium treatment records were set up, with a view to maintaining statistical information and standardising treatment. It was concluded that the total radium available from all sources in the UK was approximately 24g, which was half the amount estimated to be necessary; the only certain fact in the entire matter was that '... at the present time very little radium is being produced anywhere except in Belgium.' [7]

An editorial in *The Lancet* in 1929 on the cost of radium made two interesting observations. The first was to wonder why such expense was accepted in employing gamma rays derived from radium, when they were, in effect, similar to the X-rays available using existing, well-developed technology:

> Why pay millions for extracting a few grammes of radium from tons of foreign ore when their effect can be closely simulated by electric means when the high-tension apparatus has been brought to the necessary pitch of perfection? The reply is a simple one. The shorter the wave the greater the penetrating power. The gamma rays from radium have about a quarter the wave-length of the hardest rays yet produced, and the apparatus designed to produce these is not only very costly but undergoes rapid deterioration in use. To bring this artificial radiation up to the potency of the gamma radiation from radium would require apparatus which, if it were not beyond the wit of the electro-technician to make, would in the course of a few years cost at least as much in upkeep and replacement as the corresponding amount of radium element. For, incredible as it may appear to anyone except the hardened physicist, radium element can be employed for years without sensible loss. In 20 years the deterioration is only 0.7 per cent; radium is, in fact, an engine running continuously without working costs. [8]

The second speculation suggested, for the first time, a radical economic possibility that would be applied in later years to uranium and plutonium:

> [Radium] is in fact far and away the costliest marketable commodity. In times of civil strife and currency depreciation jewellers have been known to carry about their whole capital consisting

> of a few gems in a vest pocket, but the price of radium is above rubies. The total amount of the element ever extracted may not exceed 20 ounces, at a present money value of £6,000,000, and could (though should not) be carried in the coat pocket. There is a Radium Bank in Paris, and it would be open to a new nation to base its currency on a radium standard with the whole of its reserves in a small lead-lined safe. [9]

According to an account by two London cancer surgeons who visited hospitals and institutes in France and Belgium in the same year,[10] both these countries were well in advance of Britain in terms of organisation; certainly neither country had the same problems of supply. The Belgian operation had achieved stupendous success and the ores from Haut-Katanga sent for processing to Olen established a virtual monopoly:

> The *Bruxelles-médical* states that the factory near Olen producing radium from the Congo ore has turned out a total of 150g of radium, 4g a month on average. Our exchange adds that there are only 310g of radium known in the world to date. Certain minerals of the Congo, it is said, contain 60 per cent of the radium-bearing uranium ore, while the US mines had only 2 per cent. [11]

In common with just about every other international 'commodity', however, the acquisition of radium attracted accusations of chicanery and there continued to be allegations of manipulation of the market, just as there had been fifteen years earlier in London. This time, complaints were made to the French Academy of Sciences about a grave threat to public health:

> The radium mines of Chinkolobwe, the richest in the world today, supply an ore that is imported and the radium extracted at the works located in Olen, Belgium. It has recently become known that the management of these works have received orders from

> the owners to restrict the output in order to maintain the price of radium at its present high level. [12]

Olen's production almost certainly exceeded demand and there was probably an unwelcome choice between either placing a limit on production, or accumulating reserves whose future value could not be assured. But in just the same way that the Congolese deposits came to dominate world production after the First World War, another dramatic change was to occur: one man looking for gold and silver was about to follow a hunch.

In 1929, Gilbert LaBine, a professional mineral prospector, and owner with his brother of the Eldorado Mining Co. in Ontario, began to investigate the Canadian Northwest Territories. This vast wilderness area was known to be mineral-rich and already there had been a lucrative series of gold strikes begun after 1898 around the Great Slave Lake. Places like Yellowknife, Outpost Island and Gordon Lake joined the long list of other Canadian mineral strikes that turned wasteland settlements into centres of frantic mineral extraction – Cobalt, Sudbury, Porcupine, Red Lake and many others. The Northwest Territories would eventually develop rich gold, silver, lead, zinc, copper, tungsten and diamond mining resources, but it was LaBine's trek in 1929 that would open up the most spectacular twentieth-century prospect.

LaBine had been reading the accounts of two specialists from the Geological Survey of Canada who had first surveyed the area in 1900, and he knew he was looking for what they had described as: 'cobalt bloom on the rocks of the east shore of Great Bear Lake'. LaBine lived in Cobalt, Ontario, and knew that this 'gossan' indicated the presence of silver. He was landed by plane on the ice of the Great Bear Lake with his canoe and no more supplies than he could carry. This was an unforgiving landscape, just on the Arctic Circle, more than a hundred miles from the nearest Eskimo settlement; and it was big – the east shore of the lake was

500 miles long. LaBine identified enough potential to return to the lake the following year for an extended visit, with a colleague and 1,600lb of equipment. They built a sledge and began a trek along 200 miles of shoreline. On 16 May 1930, exploring alone, Gilbert LaBine encountered a mass of rocks bearing the cobalt bloom; there was going to be silver aplenty. As he investigated further, he realised that he was also looking at something else – a bluish-black ore. The realisation grew that he had stumbled upon a huge and potentially extremely rich pitchblende deposit. Without delay, he staked claims at Echo Bay, near to what would become LaBine Point, 26 miles south of the Arctic Circle, and arranged to have the pitchblende assayed. The excitement must have been severely tempered by the realisation of where they were – in wild, frozen ice-bound wastes, hundreds of miles from electricity or machinery; just to reach a railway would require a hazardous journey through the tortuous Mackenzie River system, much of which was frozen solid for more than half the year. In summer, sunset and sunrise were only a few minutes apart and sleep deprivation would be a major problem; in winter, daylight was fleeting and working hours would be severely reduced, even if weather conditions permitted. Before even those considerations, there was the certainty that crippling freight and transport costs would be a real disincentive to prospective investors in any plan to exploit the area.[13]

Gilbert LaBine returned again in 1931, with more equipment and more men, to continue exploration and begin rudimentary mining operations. He had already ascertained that the Belgian Union minière was not interested in assisting with the development of the resources at the Great Bear Lake; it seemed equally unlikely that Canadian finances would be readily forthcoming. However, he was convinced of the world importance of the radium that could be obtained and decided to bear the costs alone as long as was necessary to prove the value of his discovery. In the summer of 1932, he had at considerable cost assembled a small mining plant, with 200 tons of equipment brought in by train across the 300 miles from Edmonton to Waterways, by Fort McMurray in Alberta; then – while the river routes remained unfrozen – 1,400 miles by small boat, canoe and sledge using the Athabaska, Slave, Mackenzie and Bear rivers. Occasional use was made of the more direct but expensive 1,000-mile route by air from Edmonton. The settlement of Great Bear was renamed Port Radium; its small mine, saw-mill, power plant and population of 200 was serviced by a general store, post office, radio station, Royal Canadian Mounted Police post and mining claims office; there, Gilbert LaBine built himself a house, which he name Muidar (radium spelled backwards). Almost 3,000 separate claims were registered and, by the end of 1933, freight costs had reached almost $300,000 and there was no means yet identified of how or where to conduct the complex chemical processing.[14] In the comfort of their city clubs and offices, the cynics who had so confidently predicted financial ruin were rubbing their hands in anticipation. However, in May 1933, Gilbert LaBine encountered the man who would solve the chemical extraction problem and help take the entire project forward in such a manner that within a few years Canada would take over the prime position as the world's principal producer of radium – and later, uranium.

Marcel Pochon, the French-Swiss chemist who had trained with Marie Curie in the rue Lhomond, left the ailing South Terras Mine in Cornwall and joined Gilbert LaBine. The Eldorado Gold Mining Co. established a chemical processing refinery for pitchblende at Port Hope, 50 miles east of Toronto on the shores of Lake Ontario, with Marcel Pochon as director and chief chemical engineer. Port Hope was originally a tiny seventeenth-century settlement named Ganaraske, inhabited by Mississauga Indians. A Crown Patent for the township was granted in 1797 and the small town soon had a harbour, two breweries and half a dozen small distilleries, and saw its prosperity assured when the railway came in 1857. For the Eldorado company, the new radium refining plant was a seemingly forbidding 3,000 miles from Port

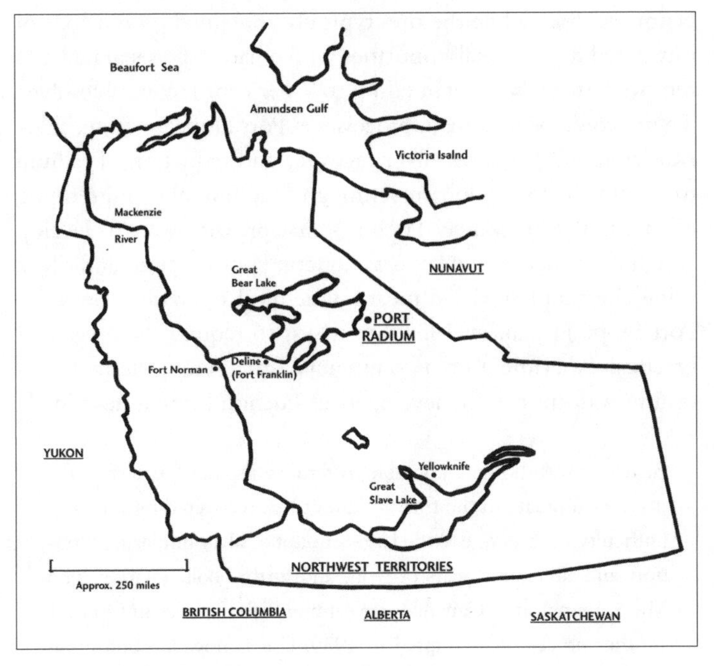

Map showing the location of Port Radium, Northwest Territories, Canada.

Radium. The difficulties appeared, if not insurmountable, financially unsustainable. The mining site at Port Radium lay in a freezing wilderness, where frost penetration reached deep beyond 300ft; the transport routes were similarly weather-affected, and the refinery on Lake Ontario had to identify and develop dozens of processing techniques to cope with the complex forms of ore expected from Great Bear Lake.

The mined ores were separated and milled at Port Radium for supply in four different grades and, although the purity of the pitchblende was high, averaging from 30 to 62 per cent of uranium oxide, there was a need to find a composite, efficient chemical process suitable for the range of ore available. The complicating factor was that, while the ores typically contained proportions of unwanted heavy metal impurities which had to be separated and removed, they also contained up to 7 per cent recoverable silver. There would be four main processes at Port Hope; roasting, chloridisation and grinding; silver recovery and reduction of radium to radium-barium sulphate; refining of radium salts; and refining of uranium salts. Marcel Pochon's first priority was to develop an efficient chemical means of extracting both radium and silver, using the simplest viable means. Chemical operations began at Port Hope in January 1933, but it was to require three years of practical experiment on a commercial scale before the best procedure was sufficiently developed, as Pochon himself described:

> At first it was thought advisable to remove the radium, letting the silver accumulate in the final residues to be recovered by smelting. Difficulty was had due to the large amount of silver hindering filtration, and after three years of work and further collaboration with Mines Branch [the Canadian government laboratories in Ottawa], another process was adopted in 1936. This is now in continuous operation at the plant. It provides for removing the silver early in the work, followed by recovery of the radium. Treatment of 150 tons of pitchblende, carrying 1,550oz of silver per ton, was completed in 1936, giving evidence that the process was economical. [15]

In addition to solving the chemical process, the Eldorado company also had to do something about the transport problem. They bought the Northern Transportation Co., which operated steel boats and barges, and owned an aeroplane for passengers and light freight which was operated by Mackenzie Air Service from Edmonton. In 1935 a large freight plane – nicknamed the 'Radium Express' – enabled 4,000lb payloads to be carried regularly between Edmonton and Port Radium (in summer using pontoons to land on water and during the much longer winter fitted with skis for landing on ice). Two supply boats, the *Radium*

King and *Radium Queen*, were brought in by crate and assembled at the lakeside, where they were used to bring in the regular supplies of diesel fuel and to take out the dozens of bags of milled concentrates en route for Port Hope on Lake Ontario.[16]

The Eldorado operations at Port Radium and Port Hope went from strength to strength. The commercial muscle of the Union minière in Belgium proved to be no barrier to its success. Indeed, the fact that the Belgian company was widely accused of being secretive and over-zealous in controlling the market and protecting its own high income probably encouraged the Canadian operation. At the time of Gilbert LaBine's pitchblende discoveries, the American Chemical Society suggested that the new source would help to alleviate the effects of Belgian exorbitance:

> The richness of this ore indicates that here is a deposit able to match itself against that from South Africa. The men who own it are primarily interested in the humanitarian use of radium. As a result of this discovery, radium may soon be available for the trained men who are skilled in its use and capable of applying it for the alleviation of human suffering. [17]

For a time, the Eldorado company claimed that it was the sole supplier of radium in the Western Hemisphere and that it was the world's largest producer. Although Port Radium was closed in 1940, it reopened the following year, with the processing of uranium attracting additional commercial interest. The same year, the company appointed the Canadian Radium and Uranium Corporation of New York City as its exclusive agent in the USA. The corporation opened a laboratory in Manhattan, where the manufacture of radium products and preparations for medical and industrial uses was concentrated.

One memoir of Port Hope, by Prof. Don Wiles of Carleton University, Ottawa, gives an account of his year spent at the refining plant from May 1947:

> When I arrived in Port Hope, the entrance to the town sported a large billboard proclaiming 'The Home of Radium'. It was mid-May, the town was very green, and so was I. But I was very enthusiastic. I learned how to handle the radium vault, and became the main employee in charge of this storage facility. The idea was to take little vials of radium, containing about 300–400mg of radium (as the anhydrous bromide) and store this in specially designed holes in lead blocks. There was a good deal of radium there at the time – perhaps several hundred Curies.
>
> My next job, and ultimately my main function, was to do a fractional crystallisation of the radium–barium concentrate, so as to produce pure radium bromide ('pure' meant greater than 80 per cent, the rest being barium). This work involved being presented every month with a flask containing about 8 Curies of radium, and working that down over the next three weeks to one or two small vials of pure radium bromide. This was essentially the same procedure as had been used by Marie Curie 45 years earlier.
>
> Safety precautions were minimal, and not well understood even by the plant medical doctor. Every Monday morning I came in and the beakers were coloured brown from the radiation. The quartz crucibles were a deep purple. These colours were removed by gentle heating. It never occurred to anyone that this same heating also removed the radon and sent it up into the air which we were breathing. We were told verbally not to do certain operations when the medical inspectors were around. That didn't matter, of course, because the inspectors never came. [18]

The biggest change in operations began during the Second World War, with new demands for uranium being made by the US government's Manhattan Project, which was producing the first nuclear weapons at Los Alamos in the New Mexico desert.[19] The demand was no longer for relatively benign end-products like luminous paint; the new frenzy was to produce fissionable uranium-235 and plutonium for nuclear weapons production. This change in the significance of mineral resources was both highly secret and highly political. It did not occur everywhere simultaneously; quite the contrary, in fact: reliable information about the expected new properties of uranium and about what different governments proposed to do with it was either unobtainable or deliberately falsified. In 1942, the US government ordered 60 tons of refined uranium oxide, and the Canadian government began secretly to take control of the Eldorado company; materials were coverty acquired for the world's first atomic weapons, assisted by the shadowy operations of a Russian-born French minerals speculator named Boris Pregel. In 1943, with the Eldorado Co. becoming a Crown corporation, the first operational nuclear reactor outside the United States was opened at Chalk River, Ontario, as an Anglo-Canadian-French collaboration. At the same time, a new two-storey radium refining plant was opened at Mount Kisco in Westchester County, New York, where radium and other nuclides were extracted from residues supplied from Port Radium and Port Hope. These activities continued until the late 1950s, when the Mount Kisco plant was ordered to be closed for non-compliance with legal employee exposure limits and the entire radium operation was moved back to Manhattan.

By the mid-1950s, there were over 300 areas in Canada alone which had been identified as having potential for uranium mining. Most of these sites lay along the 100-mile-wide edge of the geological 'shield' from the east of the Great Bear Lake to Lake Huron.[20] During the second uranium period, when the initial military secrecy had subsided and the principal uses for uranium were in electricity generation, Eldorado Gold Mines became Eldorado Nuclear Ltd in 1966. Port Hope still specialises in processing purified uranium trioxide 'yellowcake' in liquid form (known as 'OK Liquor') from the company's plant at Blind River, on Lake Huron, Ontario. (This location is near Elliot Lake, which was developed after the Ontario government had opened 100,000 acres of land for uranium prospecting in 1968; there were 2,500 claims, chased by a thousand prospectors, and the Elliot Lake site became for a time the world's biggest uranium mine.) The OK Liquor is converted into uranium hexafluoride gas – a highly corrosive compound of uranium and fluorine used in separating fissile uranium-235. Eldorado was further commercialised in 1988 when the assets were combined with the Sasketchewan Mining Development Corporation to form the Cameco Corporation, which has extensive uranium and gold-mining interests.

Today, Canada, with 32 per cent, is the world's largest producer of uranium, all of it coming from Saskatchewan, where ore quality is very rich. Today, the 'richest uranium mine on Earth' is at the Key Lake Mine in northern Saskatchewan, where background radiation levels at the open pit are 7,000 times greater than 'normal'. This is the most productive uranium mine in the world, where the nearby mill produces the powdered, concentrated 'yellowcake'. In the province, a total of 10,463 tons of uranium ores produced 27.2 million pounds of U_3O_8 in 2003 alone, and known deposits are expected to last for at least twenty-five years at current rates of extraction, regardless of the prospects for the discovery of further resources.[21] World uranium ore deposits (as distinct from production) by country and percentage are as follows:

Australia	28%
Kazakhstan	20%
Canada	14%
South Africa	10%
Brazil	8%
Namibia	8%
Russia	6%
Uzbekistan	4%
Mongolia	3%
USA	3%
Niger	2%
Ukraine	2% [22]

With interest in radium fading in favour of the frenzy for uranium, and the focus moving from medicine to weaponry, the world changed forever. In Belgium, Union minière maintained a strong position against Canadian interests, despite continual accusations of making grossly excessive profits at the expense of cancer sufferers. In the UK in particular, there was righteous anger at the Belgian pricing policy, and during 1929–30 there were moves in the British House of Commons to involve the Health Department of the League of Nations in Geneva. The British Minister of Public Health observed that, 'that ought to bear upon the Belgian producers in order to bring them to reason'.[23] It is not clear whether the League of Nations took action, but at any rate (probably due to activities in Canada) Belgian prices were reduced in the 1930s.[24] Olen also undertook considerable uranium production and, although information is scarce, it is known that the US government made substantial purchases of Belgian uranium. The powerful head of Union minière for many years, Edgar Sengier, was involved in the highest-level international political negotiations in the lead-up to the Second World War. He fled to New York to avoid the German invasion of Belgium, and secretly arranged the shipment of 1,200 tons of extremely rich pitchblende, which was mysteriously cleared through US Customs and stored in a Staten Island warehouse. It was not called upon for the Manhattan Project for another three years – a delay that gives credence to the claim that Sengier had concluded a secret deal with the US government. Moreover, it seems to have been an exclusive arrangement, for when Sir Henry Tizard tried to negotiate with him on behalf of the British government, Sengier rebuffed him, despite the fact that the Belgian government-in-exile would be established in London. He did, however, agree in principle to a French proposal to collaborate in building an experimental atomic weapon in the Sahara Desert – a project only thwarted by the outbreak of war.[25]

Ore from Chinkolobwe was transferred to Germany, although some was moved or hidden in attempts to prevent its forfeiture; some went to France, some to Morocco and 8 tons of uranium oxide was hidden in a cellar in Delft during the German occupation of Holland.[26] In 1944, the US established the anodyne-sounding Combined Development Trust, the purpose of which was to monopolise control of the world's uranium resources. The trust reported that the US and the UK would together control over 90 per cent of the world supply if they gained exclusive access to the output of Chinkolobwe. As soon as the war ended the following year, the Belgians agreed and, although the uranium went to the USA, half of the costs were paid by the UK.[27] In all, about 4,000 tons of high-grade ore reached the USA (excluding two shipments that were sunk by German U-boats). Sengier was given the US Medal for Merit for his role in supplying uranium materials to the USA during and after the Second World War. (He later angrily returned the medal to the White House, appalled at the behaviour of the USA and the United Nations in failing to protect the people of the Belgian Congo in the bloodbath of 1959–60.)

Although it was known by the 1950s that the production site at Olen was dangerously polluted, as were similar facilities elsewhere, the plant continued producing radium and uranium materials, and a number of precious metals, until 1980. In 2001, Union minière du Haut-Katanga became known as Umicore and the company, which remains a major minerals and metals producer, has new interests in materials recycling, environmental management and sustainable development. The post-war years saw the beginning of the first world-wide uranium period; during this time, nuclear technology developed rapidly and the numbers of civil and commercial nuclear reactors increased dramatically. New radionuclides with safer, purpose-designed properties and neutron-beam technologies were developed for new forms of nuclear medicine. These developments finally brought to an end the extensive use of radium in medicine and industry, although many years passed before the final disposal of stocks of countless, by then unreliable, radium products. The developing technology, particularly in its military guise, was a fiendishly complex, secret and energy-intensive process and the separation of fissile uranium-235 from the non-fissile uranium-238 – the holy grail of the entire post-war period – diverted vast national resources throughout North America and both Western and Eastern Europe for decades.

At the end of the war, as official interest in discovering uranium minerals increased, frenzied activity broke out across Canada, Australia and the USA in particular, as prospectors professional and amateur took up the call, just as men had always done in the past over other mineral resources. One dash for uranium shares on the Toronto Stock Exchange resulted in claims being staked in the grounds of a church at North Bay, Ontario. In another area, 10,000 separate claims were registered in a few weeks and new mining towns such as Uranium City in Saskatchewan sprang up in the most unlikely landscapes.[28] Here, in the Beaverlodge area on Lake Athabaska, a new uranium refining plant was opened by the Eldorado Co. in 1953.

In the USA a government order of 1943 prohibited the exploitation of uranium for any purpose other than military, and geologists and spies were sent out to report on the potential uranium resources of over fifty countries. At home, such as at Central City, Colorado, a few miles from Denver, abandoned workings, which had in the past produced fabulously rich gold and silver ores, were successfully reopened in the largely unsupervised search for uranium. The new searches were directed at the unbridled patriotic fervour of the majority of Americans and were shamelessly promoted by the US Atomic Energy Commission:

> The security of the free world may depend on such a simple thing as people keeping their eyes open. Every American oil man looking for 'black gold' in a foreign jungle is derelict in his duty to his country if he hasn't at least mastered the basic information on the geology of uranium. And the same applies to every mountain climber, every big game hunter, and, for that matter, every butterfly catcher.[29]

Americans reacted with enthusiasm, buying 35,000 mail-order Geiger counters in the year 1953 alone and, with gas stations offering free 'uranium maps', uranium-hunting picnic trips became a favourite weekend leisure activity. Popular magazines and newspapers promised headlines such as, 'Get rich from that miracle atom' followed by pages of features on 'How uranium makes new millionaires overnight'.[30] Indigenous Indians, who had in previous years been fobbed off with supposedly worthless land by the government, found that they were adept at identifying ores on their homelands that, ironically, proved to be well endowed with the newly favoured minerals. Particularly in the mineral-rich states of Colorado and Utah – where fundamentalist polygamists were a potent social and political force – shares in real or imaginary 'strikes' were aggressively bought and sold. Sometimes, share certificates were traded or even given away in exchange for the purchase of hamburgers and toothpaste – even bar-room hookers added that trick to their usual offers. Geiger-counter kits could be obtained in exchange for cereal-packet tops, and bars and diners across Colorado displayed intimidating signs demanding 'no talk under $1,000,000'. Everyone became an expert, and there were 'How to do it' books in every second store. Most were ill-informed and badly written; others, such as the invitingly titled 'Uranium – where it is and how to find it', were sensible, valuable guides written by professional geologists. As its authors stated on the first page:

> Neither the Gold Rush of '49, the diamond strike of South Africa, nor the Klondyke gold stampede of '98 can compare with the intense earth-encompassing search now underway for uranium minerals. [31]

Inevitably, there were stories of intrigue, deception, theft and murder. And to fuel the frenzy, there was a classic success story. Charlie Steen, a life-long well-travelled professional prospector, was down on his luck after returning from South America when, with his wife lying ill in their $15-a-month shack and his last drill-bit on the point of shattering, he struck a fabulously rich source of uranium ore on the Colorado Plateau of eastern Utah. Within a year, estimates of the value of the uranium and vanadium in his claims ranged from$10 million to $300 million. His success bred even more success, as he invested in dozens of uranium and silver-mining operations, airlines, newspapers and utilities. After ten years of enormous profits, when he sold his initial mine at Moab, the price was $25 million. Although he became fabulously wealthy, in an uncanny near-echo of the life of the legendary Horace Tabor, who had made an extraordinary fortune in Colorado silver in the 1860s, Steen quickly lost everything and only narrowly avoided being declared bankrupt.

But uranium hysteria was rampant; during one month in 1954, when there were over 500 uranium companies listed for the Colorado Plateau (each potentially owning hundreds of claims), 30 million uranium-mining shares were traded in Salt Lake City alone. As with all such mineral booms, most of these shares were worthless, but some would turn out to be fabulously valuable. The hoardings declaring, 'Uranium – creating new men of wealth, uraniumaires!' acted as recruiting sergeants in the frantic drive to motivate ordinary citizens who knew nothing of, and cared less for, such complications as alpha particles, gamma rays and radon daughters. For one 'uraniumaire' who had previously suffered from poor television reception, the main benefit to human life was the installation of a television set in his private plane and the ability to instruct his pilot to take to the air whenever he wanted to see the *Mickey Mouse Club Show*.[32] Such frivolities were soon swept aside, however, as the big mining and energy corporations moved in to dominate the uranium business. The AEC, often accused of deceitful below-the-belt tactics, called a halt to the uranium prospecting boom in 1958; however, this was seen as a mistake, as energy producers abandoned fossil fuels and turned to nuclear generation, and the freelance uranium hunters were welcomed again after 1965.

The state joined partnerships with the big corporations, and secret chemical extraction plants, military nuclear reactors and weapons development, assembly and testing areas appeared across the USA. The Cold War was made manifest.

In the USSR, rich sources of pitchblende had been discovered in the Fergana Valley in Uzbekistan and had been extracted by a private company between 1910 and 1914, after which the operation was overrun by the Red Army. A State Radium Institute was established in Leningrad (the ancient capital of Russia, now once again St Petersburg) in 1922, and others later appeared in Kharkov and Sverdlovsk, where work on radium (and later on fission theory) was conducted. In 1945, Joachimsthal fell to Soviet occupation (along with, incidentally, areas of North Korea that had exploitable uranium sources). Lavrenti Beria, Stalin's chief of secret police, was put in charge of Soviet nuclear weapons operations. An agreement was quickly concluded between the USSR and the Czech government for the supply of uranium ores and the mines at Joachimsthal were restarted, with Czechoslovakia permitted to retain any radium available after everything else had been delivered to Russia. Other Czech resources were similarly plundered and in that country alone seventeen forced-labour camps were established to extract uranium ore[33]; similar slave-labour mining was organised across the Soviet empire. France and China were the remaining countries that set out on the long uranium–plutonium weapons trail in the immediate aftermath of the Second World War.

As the military, professional geologists, obsessed patriots and spies re-examined half of the globe in the search for uranium ores, the sites of earlier radium operations were subjected to equally secretive probing, in the hope that sufficient uranium materials might be encountered. Even in Britain, at least two of the earlier radium processing sites were inspected by government departments. During the summer of 1948, a local man who with his brother had worked as a labourer at John MacArthur's Loch Lomond Radium Works, reported a stranger at the site (which was then, as today, used as a boatyard). The result was that the existence of an abandoned radium plant was brought to the attention of a hitherto ignorant government which had by then embarked on its own nuclear weapons programme and was searching for uranium ore wherever it could be found. It was decided that the site at Balloch was of interest – not for reasons of public protection, but as a potential source of uranium ore. Under the protection of the chief constable, a secret survey of the site was carried out for the Atomic Energy Secretariat of the Ministry of Supply. The survey estimated that the site comprised 200–300 tons of sand and slime residues and 20–50 tons of untreated ore. Three pits were dug and analysed and a wide range of radiation readings was encountered throughout the site. It was found impossible to understand the nature of the site, which was regarded as ' . . . far out of radioactive equilibrium':

> It is not possible to determine the precise amount of available uranium by a simple Geiger Counter survey. Such a determination would require close trenching and sampling of the whole site to obtain a large number of specimens for chemical analysis. On present evidence the total uranium present seems unlikely to exceed a few hundredweights, only a small part of which would be extractable; and undoubtedly the cost of detailed survey and of chemical analyses would exceed the value of any uranium we might recover. [34]

Public safety issues were not considered and there were no recommendations for decontamination of the site or protection of the public. No action was taken and the Loch Lomond Radium Works played no further part in Britain's atomic bomb programme. The same department had already examined the records of the South Terras site in Cornwall and, four years later, the Geological Survey conducted test drilling there, with unknown results. However, the search for uranium in Britain continued for many years, to no effective outcome. In 1957–58 the Department of

Scientific and Industrial Research conducted aerial radiometric surveys of several areas of Cornwall and Scotland, as a result of which abandoned mine workings were reopened and trenching and exploratory drilling undertaken.

In thirty years, the tiny industry that promised so much for medicine and had simultaneously excited the public with promises of luminous watches and light-switches had changed out of all recognition. Medical benefits would continue and become more sophisticated, but the intervention of military interests and funding, and the Cold War atmosphere of international political suspicion, turned what had been thought of as a rather quirky, frivolous discovery into the huge, inhumane and socially destructive technology that has since had such an over-bearing effect on human existence. The nuclear industry will never go away and its significance can hardly be overstated. This seemed to be recognised when Glenn Seaborg (1912–99), who first identified plutonium (and eight other elements produced by neutron bombardment of uranium), speaking of plutonium, repeated in 1970*4 the interesting, if curious, idea that had first surfaced in relation to radium in 1929 and had been repeated about uranium in 1947:

> Breeder reactors will be the backbone of an emerging nuclear economy, and plutonium will be a logical contender to replace gold as the standard of our monetary system. [35]

Not all informed opinion agreed. One of Ernest Rutherford's star young physicists, (Sir) Mark Oliphant (who made the first attempt in 1925 to transmute uranium, discovered tritium in 1934, and worked on the Manhattan Project in the USA) became a vehement opponent of nuclear weaponry, concluding that, 'nothing can wipe away the poison uranium has brought to mankind's affairs'.[36] Today, the entire nuclear industry's controversial future is severely compromised by its inability to solve a major issue that began in the earliest days of radium – the deadly legacy.

ELEVEN

LEGACY

The legacy left to us by radium and uranium (and the other radioactive elements) is related to their spontaneous emission of energy, the kinetics of which is characterised by the radioactive 'half-life' of each nuclide. This is the period of time taken by the nuclei of each to lose half of their activity through radioactive decay to the 'daughters' in the chain (which in these natural series are themselves usually radioactive). As Appendix II shows, each successive daughter in a decay chain does not necessarily have a shorter half-life; unsupported elements with a long half-life will be with us emitting radioactivity for longer than those with a shorter half-life and daughter nuclides with short half-lives can be supported over long time periods by their longer-lived parents; the quantifying of resulting health effects is, inevitably, a very complicated affair.

In the case of radium-226, the half-life is 1,602 years; naturally occurring uranium-238 has a half-life of 4,500 million years. All the related materials, residues, products, buildings and production sites connected with these radioactive elements are subject to the same implications of that time factor. This – as it applies to waste materials – is the huge unresolved issue for the nuclear industry. Some analysts insist that the technical and financial implications are so staggeringly vast and long-term as to make any commercial use of nuclear power wholly and indisputably uneconomic; any suggestion of continuing to promote nuclear technology without an economic solution to the waste problem is regarded as unacceptable and unsustainable. Military use of nuclear materials is another matter again. This chapter will reveal some of the problems that stemmed from the 'radium days' but cannot altogether ignore the issues that arose from the subsequent world-wide desperation to mine uranium-bearing minerals; in particular, it will recount some of the problems that have affected the sites and buildings that have featured in earlier chapters. Little attempt will be made to cover uranium-mining operations over the last fifty years, or military activities and their extraordinarily extensive and costly implications. (In the USA, for example, the Department of Energy controls over 20,000 buildings on more than 2,000 sites which could cost an estimated $500 billion over 75 years to decontaminate.)[1]

Radioactivity cannot be 'neutralised' or destroyed and attempts to 'sweep it under the carpet' are worse than pointless. Particular features of the problem as it relates to radium are that during much of the period when radium and radon were in production and use, the methods employed were relatively crude, few if any records were maintained and there was scant regulation governing the industry. Much of what took place has been forgotten and often attempts to redevelop land fall foul of these earlier activities. There is no such place as 'away' in which to throw or dispose of radioactive materials. Even when some of the early production sites have been identified and decontaminated, the methods used raise questions of their own. The use of radium, emitting intense gamma radiation, is almost unheard-of today, as short-lived man-made radioisotopes manufactured in particle accelerators and nuclear reactors can be safely and precisely tailored to particular uses. But the after-effects of many years of extracting radium and its use in an enormous range of devices and products are still with us. Radiation can be buried, but never ignored; out of sight and out of mind, it can leach and migrate and contaminate in ways that may not be controllable or even identifiable. Modern luminous materials use aluminium oxides and rare earth ceramics and the radionuclides which have replaced radium do not emit high-energy gamma radiation. However, to achieve the degree of luminosity given by radium requires substantial quantities of nuclides such as tritium and promethium. When millions of modern watches, telephone dials, smoke detectors and the like are disposed of, the resultant radiation waste problem could be significant if – as often happens – appropriate procedures are disregarded. In recent years, even prestigious medical and academic institutions, including Oxford University, have been accused of incompetent and illegal flushing of diagnostic and other radioactive materials into domestic drainage systems.[2]

The legendary Bohemian pitchblende mine and refining plant at St Joachimstal, where the death-rate among miners was thought to have been substantial, were abandoned in 1940 and turned into a public park. Widespread radon concentrations were partly due to medieval silver-mining activities conducted in underfloor domestic cellars, where disused adits often interrupted pitchblende deposits. However, early in the Cold War, the Soviet Union reopened the entire area for uranium mining and, because of the secretive conditions that prevailed, little of what took place there can be fully understood. The Erzgebirge mines employed 100,000 men in the forced-labour extraction of uranium for Soviet nuclear arms manufacture until the early 1960s. People who were imprisoned without trial for unspecified breaches of 'loyalty' referred to 'Jáchymov Hell'.[3] Roads and buildings were commonly constructed of uranium waste and in 1980 radon levels were encountered which reached 500 times the maximum level accepted in the UK beyond which remedial action is required. The cost of sealing one mine alone was estimated at £150 million.

In 1989, the Czech government refused pleas for evacuation from people living near the site of the mines. Although the site was only three miles from the border with (then) East Germany and thirty miles from West Germany, the Czech government was desperate not to jeopardise its own nuclear power programme.

Jáchymov lies in the heart of a popular tourist and spa area at Karlovy Vary (formerly Carlsbad) and the lure of foreign currency – and a desire to return to the good old days when European aristocracy sipped spa water – could not easily be reconciled with the admission of such an environmental problem, far less its treatment. Today the many establishments based around the opulent Radium Palace Hotel boast a spa company with a thirteen-page price list covering a huge range of services and treatments.[4] The radon-rich waters within the town originate in what was one of the oldest mines in Europe. The Svornost silver-mining pit was established in 1518 and subsequently became a source of uranium minerals. The first radon spa-house opened in 1906, using water brought from the pit in buckets by teams of horses; today sophisticated systems of pipes distribute the water throughout the town. After thirteen years of dispute, agreements between the governments of Germany and the state of Saxony envisage spending €78 million on the reclamation of many of the uranium-mining sites until the year 2012; this, however, represents only about 17 per cent of the total amount required.

Other uranium-mining areas in Czechoslovakia, Poland and East Germany fell under Soviet control after the Second World War; one uranium mine alone in Poland had 26,000 workers. In Russia itself, huge areas have become notorious as the devastated legacy of the Soviet bomb-making project. In Central Asia, in the southern Urals at Mayak, near the city of Ozyorsk in the Chelyabinsk region, there are millions of tons of uranium wastes and widespread poisoning of water-courses. Conditions in such vast work-camps were known to be appalling, with little consideration given to the well-being of the workers. With regimes which for many years did not encourage medical complaints or employ health monitoring, and with the unrecorded dispersal of the workforce following closure, it has been virtually impossible to compile any coherent epidemiological evidence.

In the former East Germany, uranium mining after the Second World War came under the control of the Soviet Wismut Co.. With a workforce totalling 150,000 at any one time and managed by 5,000 Soviet employees, it produced 220,000 tons of uranium for shipment to the Soviet Union from 1945 until mining ceased in 1989 with the unification of Germany.[5] The secretive company (its name means 'bismuth', no doubt for purposes of deceit) operated as a state within a state. Today, derelict uranium mills with hundreds of millions of tons of residues and mine tailings covering 1,000 square kilometres dangerously pollute the devastated landscape and the extent of the problem is appalling. It is estimated that 1,600km of mine workings up to 2,000m deep will require filling or flooding; 300 tunnels, 85 ventilation shafts and dozens of waste ponds containing radioactive sulphuric acid will have to be decontaminated or reclaimed, with particular attention given to the protection of groundwater supplies. Pilot reclamation projects in such areas have been completed under European Union programmes and the reorganised Wismut Co. now undertakes reclamation works estimated to cost at least $10 billion across eastern Europe.[6]

The medieval 'Bergsucht' visited upon the early miners of Saxony and Bohemia by the malevolent dwarves guarding the mines was nothing compared to the suffering experienced by the hundreds of thousands of slave labourers of twentieth-century Eastern Europe. The work begun in the mid-1920s by Harrison Martland and taken up by later epidemiologists and biologists led over the decades to regular strengthening of the international regulations for radiation workers, and for miners and others exposed at work to radioactive materials, as well as for medical patients and members of the public. Despite these advances, the adherents of radiation hormesis claim that dangers of exposure to low-level radiation are unproven; indeed they insist that such exposure may be beneficial, since organisms have inherent abilities to detect and repair damage. However, it is now generally accepted that any exposure to radiation carries risk. Consequently, there is concern that there should be no unnecessary exposure to man-made radiation. The International Commission on Radiological Protection insists that: 'No practice involving exposures to radiation should be adopted unless it produces sufficient benefit to the exposed individuals or to society to offset the radiation detriment it causes.'[7] This is a good first principle, despite the rather unconvincing implication that benefit for the individual and for society necessarily coincide.

It is an irony that, during decades when conditions were being slowly improved for workers in uranium mines, there was a simultaneous move for the unregulated exploitation of abandoned mines in 'health treatments'. Even today, there are many adherents of unsupervised radon inhalation and radon balneology and the practices have generated as substantial an international business as did the drinking of radioactive water by the likes of Eben Byers decades before. In the Soviet Union during the 1980s, the national health system was prescribing 25,000 radon baths every day and an automated factory in Moscow was daily producing 9,000 high-level radon treatments; today there are 3,500 spas and 5,000 'state rehabilitation centres', an unknown number of which offer radon treatments. The Czech Republic has 52 mineral water spas and 1,900 mineral springs and Japan has 1,500 spas, with the Misasa and Tamagawa Spas in particular offering radon therapy. Other spas specialising in radon are Lake Heviz in Hungary and Bad Kreuznach, Fürstenzeche, Sibyllenbad and Schlema in Germany (where, in a touching and populist enthusiasm for 'environmental improvement', a golf course is to be built on the reclaimed 57-hectare waste-heap of the former uranium mine).

At the ancient spa town of Badgastein, near Salzburg, where gold was mined in pre-Christian times, the industry enjoyed particular prosperity between the thirteenth and sixteenth centuries. The town had also enjoyed success as a thermal spa since the seventh century, with as many as nineteen hot-water springs in the centre of the town; eventually, over 100 hotels catered for the self-indulgent who roamed Europe unloading their money on all manner of frivolities. As late as 1940, when attempts were begun by the Third Reich to reopen the gold mines, what they found was not gold but a huge cavern with a very high radon content, high temperature and high humidity. It was noted that the wartime conscript miners (all otherwise unfit for military service) had begun to exhibit improved health (the Eben Beyers effect?) despite enduring a 60°C temperature difference between the tunnels and the outside air. The temperature was much higher than geological conditions indicated and research previously conducted by Marie Curie showed that the town's thermal waters had temperatures of 48°C, with a mean radon content of 800 becquerels per litre in ambient atmospheres of 99 per cent humidity.

After the war, the Radhausberg–Unterbau–Stollen tunnel, which had survived wartime orders for its destruction, was reopened. The Heilstollen, or Thermal Tunnel, was opened in 1946 complete with small cars to carry bed-ridden patients to the radon-rich galleries, where they could breathe in the 'health-giving' radiation. By 1980, it was reported that a million night lodgings were spent at Badgastein.[8] Recent tourist information enticed prospective visitors with the promise that:

> Good health comes from the Gastein Healing Gallery, deep in the core of the Radhaus mountain. The tunnel train leads you along the tracks of the old gold diggers. At that time, the healing power of the mountain was discovered. Miners, afflicted with rheumatism, were healed of their troubles. The healthy, too, recognised

> their increased capacities for performance. This healing effect is available to you in its entirety. The Healing Gallery with its combination of radon inhalation, high temperature and humidity, unique in the world, is the strongest and most effective treatment agent in Badgastein.

Imitating the advertising of some of the worst of the medical charlatans of the 1920s, the conditions were listed for which the Healing Gallery offered improvement:

> Inflammatory rheumatism; Bechterew's disease; arthroses; asthma (bronchial asthma); damage to the spinal column and ligament discs; inflammatory nerves; sciatica; scleroderma; paralysis and functional disturbances after injuries; circulatory problems of the arteries, smoker's leg, diabetes, arteriosclerosis; problems with venous blood circulation; heart attack risk factors (high blood pressure, over-weight, lack of exercise, post-heart-attack cure (six months)); infertility problems and premature ageing; potency disturbances; urinary tract, gout and suffering due to stones; paradontosis. [9]

The Gastein Healing Gallery suggests that a typical course of treatment would double the normal annual exposure from natural background radiation; a four-week cure, with twelve entries to the gallery costs €577 (accommodation and subsistence excluded). The establishment is recognised by German and Austrian national health insurance administrations and treatments are administered under medical advice (unlike most similar establishments, which specifically declare themselves not to be a medical service).[10] Quite how the passive unsupervised inhalation of a radioactive gas will counteract the effects of lack of exercise is entirely unproven. Despite all the prior evidence against such practices, some recent reports have claimed that treatment in the Thermal Gallery appears to confer an analgesic effect and an increased pain threshold lasting up to six months. The peculiar appeal of mystery and magic seem to be at work again. In the same caves where miners had to breathe through fine lace and from which they eventually fled, fearing the deadly work of evil dwarves, wealthy tourists wearing designer beach clothes now lounge around against the rock faces, paying heavily for the presumed privilege.

In the USA too, there are scores of abandoned uranium mines scattered over a dozen states which require remediation. The craze for radium/uranium cures flourishes in places which cannot boast the history of, for example, the town of Hot Springs, Arkansas, which still benefits somewhat from the discovery of local radon many years after its original establishment as a spa in 1832. There are towns named 'Radium' (sometimes seeking nothing more than a 'vogue' name) in the states of Virginia, Texas, Minnesota, Kansas and Colorado; 'Radium Springs' in New Mexico, Georgia and Arizona; and perhaps the most famous, Radium Hot Springs in the Canadian province of British Columbia. Near the towns of Boulder and Basin in the mountainous mine country of Montana, there are half a dozen defunct mines which now attract tourists by offering 'healing galleries' and radon-water. With radon levels nearly 200 times federal limits for houses, derelict mines with names such as the Sunshine Health Mine, Earth Angel, the Merry Widow and the Free Enterprise ply their doubtful trade. The Sunshine has a list of treatable conditions almost as long as that of Badgastein; its near-neighbour, the Merry Widow Health Mine, owned by Miracle Mines Inc., encourages visitors to breathe the radon for four hours a day over ten or eleven days at a time for $3.50 an hour. Photographs show elderly people, crippled from years of chronic disease, queuing at the mine entrance, having made their way 1,000ft up the hillside. Visitors are shown being carried into the mine by friends on stretchers, or on wagons abandoned by the mining company. They sit on benches in the gloom, breathing deep of the 'miracle' radon. At least one establishment, in an abandoned store in Lone Rock, Wisconsin, can only claim its 'efficacy' since the day its owner lined the walls with several thousand pounds of uranium wastes imported in sacks from Nevada.

The American fad for radon healing began by accident at the Free Enterprise Uranium Mine in Boulder, Colorado, but today the establishment boasts an extensive website complete with anecdotal, medical and scientific testimonials (although it carefully points out that, while enquiries from doctors are welcomed, 'this is not a medical facility'[11]). The recommended exposure time per course is 32 hours over several days, at a cost of $150 (in 2002). The 400ft-long 'therapy gallery' is served by an Otis elevator, lighting, heat lamps and padded benches and, for the truly enthusiastic, there is a dedicated pet-therapy service. The success of radon mines as health farms became popular following a report in *Life* magazine in 1952, which quoted with gloomy resignation the opinion of medical authorities that 'Barnum was right'.[12] Phineas Barnum was the flamboyant showman who began his career as a deceiver by exploiting human freaks, including 'General Tom Thumb', and went on to found the famous Barnum & Bailey circus. To him is attributed the adage, 'There's a sucker born every minute.' He died a multi-millionaire in 1891.

The enormous pitchblende mine at Chinkolobwe, operated initially for radium by Union minière du Haut-Katanga, ceased being profitable in the uranium market in 1960 and was closed; amid concern for safety, the company filled the huge main shaft with concrete. In recent years, however, the increasingly lawless country has witnessed a proliferation of illegal mining activity. Since 1998, 6,000 men have plundered the Chinkolobwe site every day, ostensibly in search of lucrative cobalt minerals to sell on the black market. However, there has been international concern that considerable quantities of uranium have been removed from the dangerously contaminated area, with the possibility of uncontrolled distribution and use. The government of the Democratic Republic of Congo has been under pressure to regain control of the site and in January 2004 President Joseph Kabila ordered the site to be closed. Reports suggest that this has had no effect and, at the time of writing, it appears that the former uranium mine continues to be plundered with uninhibited vigour.[13]

Union minière's radium operations at Olen near Antwerp began to decline in the early 1950s, when new, safer nuclides were developed. In previous years, as the result of regular expansion of the plant, a local watercourse known as the Bankloop was regularly diverted, resulting in widespread distribution of contamination. Apart from radioactive contamination of water and soil, there remain in storage considerable amounts of finished radium products and many thousands of tons of residues in varying degrees of radiological stability. Production finally ceased in 1980, and in 1993 Union minière (and its successor in 2001, Umicore) began a series of consultations with regulatory agencies and the local community.[14] Umicore is required to decontaminate not only the land which it owns, but land, watercourses and streets in the vicinity. The entire operation is extremely complex and work on the ground is not expected to begin until 2006 at the earliest.

In Canada, a federal research team was sent from Montréal in 1945 to examine Port Radium on Great Bear Lake in the Northwest Territories. They found the conditions to be appalling. Despite some improvements, concerns remained and in 1949 the US Atomic Energy Commission, which at the time was the sole purchaser of ore from Port Radium, cynically insisted that no information about the mine should be published.[15] Several years of deceit and secrecy about the state of Port Radium and the health of its workers followed before the mine was closed in 1960. Twenty years later, studies were begun into the incidence of cancer deaths among former Port Radium workers. Today, the Dene, one of the First Nation peoples who live around the shores of Great Bear Lake and who provided a large part of the Port Radium workforce, are at the heart of attempts to achieve assistance, environmental and social assessment, and complete clean-up and monitoring of their lands and society. The Dene, whose main township is at Deline (formerly Fort Franklin), are surrounded by millions of tons of mine tailings in the immediate vicinity and along the original ore transportation routes; they eat fish caught in the lake where vast quantities of waste were dumped and, as miners or 'radium coolies', the workers were in constant intimate contact with the pitchblende.

Problems at the other end of the Eldorado operation in Ontario were worse. From 1933 to 1939, waste materials were disposed of within the Port Hope plant; until after the war, further disposal took place at several unofficial sites within and around the town. Contamination was widespread; there was spillage during shipment, unmonitored and unauthorised dumping and recycling of rubble from older Eldorado buildings as facilities were extended or demolished. Uranium wastes were even sold to the public for a variety of building uses, spreading contamination ever wider. The total volume of contaminated material amounts to about 1 million cu.m.[16] Following a warning in 1967 of high radium contamination, the first comprehensive report by the Canadian Atomic Energy Control Board *5 in 1976 (when radon examination was made of every building in Port Hope) identified forty-seven sites within the town that needed to be tackled immediately; some sites had surface-level contamination 1,000 times normal background.[17] However, it was only in 2001 that agreement was reached between the municipality of Port Hope and the Canadian government defining the terms under which waste management and remediation will be carried out.

Largely due to widespread dumping of waste materials in what were at the time undeveloped areas, the activities begun by Sabin von Sochocky and extended by the US Radium Corporation resulted in considerable problems in the linked New Jersey townships of Montclair, West Orange and Glen Ridge. In total, an area of approximately 210 acres has been designated as contaminated, along with 440 out of 747 homes, and a number of buildings which were constructed of contaminated material; there were also a number of related groundwater problems. As happened elsewhere, considerable volumes of wastes were used as infill or were mixed with concrete for pavements and foundations. The US Agency for Toxic Substances and Disease Registry has produced detailed assessments, in association with the Centres for Disease Control, the US Environmental Protection Agency and local health authorities. Work began in 1983 to assess and quantify the problem and to design appropriate means of remedy. A number of homes were demolished, others fitted with ventilation systems to reduce radon concentrations and some had gamma radiation shielding installed. Most of a total of 250,000cu.m of contaminated soil has been removed and disposed of as radioactive waste at a dedicated site in Utah; much of this work involved reconstruction and diversion of underground utilities such as water, gas and sewerage.[18] A former toy factory site between Bloomsburg and Berwick, Pennsylvania, operated from 1948 by the US Radium Corporation, and since 1980 by the Safety Light Corporation, has been contaminated by nearly a score of radionuclides, and the nearby Susquehanna River has been significantly polluted by leaching and overspill from waste lagoons. Despite pressure from the state environmental protection agency, federal funding for decontamination in this case has been refused.[19]

In Ottawa, Illinois, the Radium Dial Co. building was used for many years as a meat store and, according to testimony given by people who worked there, a high percentage of employees suffered and died from cancer.[20] The building was demolished in 1969, but the site was still contaminated. The radioactive building rubble was dumped on various sites, including one which became a football pitch. Benches and other fittings belonging to the company found their way into homes in the town – still radioactive. Despite much anecdotal evidence, the Argonne Laboratory has not found detectable body burdens of radium in any test subject who had worked in contaminated buildings or who had been exposed to tailings sites. However, an aerial survey indicated thirteen areas around Ottawa with abnormally high levels of gamma radiation, suggesting widespread uncontrolled dumping and recycling of materials. The Luminous Process Co. was forced to cease trading in 1977 after repeated violations of health regulations. The buildings were pulled down by the Illinois Department of Nuclear Safety between 1984 and 1986 after intense public concern, but demolition workers, who wore no protective clothing, were compelled to sign waivers accepting personal liability for any illness or injury. It has since been alleged that the town's watercourses have been contaminated by radiation leaching from the site. This contamination may be due to the fact that deep well waters in areas of northern Illinois and eastern Iowa have a naturally high radon content; however, in 1991 Ottawa was included in the US Environmental Protection Agency's National Priorities List.[21] There are similar problems in other parts of America, some as a result of commercial dial-painting industries in such places as Athens, Georgia, and in Michigan, at Benton Harbor, Belding, and Bear Lake.

Ten US states that had been involved in milling uranium ores have benefited from the Uranium Mill Tailings Remedial Action Project and twenty-four designated sites have been decontaminated. In Canonsburg, Pennsylvania, about 20 miles from Pittsburgh, the radium extraction plant operated by the Standard Chemical Co. sold a single gram of radium to Marie Curie in 1921 for $120,000. From 1942 until 1960 the 18-acre plant, then extracting uranium for Vitro Metals, also supplied the town's residents with radioactive residues for use in gardens and buildings. Soon after the plant's closure, it was decided that the thousands of tons of tailings were indeed radioactive and should be dumped. A huge hole was dug, the waste material shoved in and the site was developed as a play park for children. Twenty years later, after unexplained deaths in the community, it was admitted that the material was still highly radioactive. Some buildings were demolished, 160 others decontaminated and 250,000 tons of debris was sealed at a cost of $20 million.[22] In the Colorado township of Grand Junction, the random 'recycling' of residues from the Climax Uranium Co. had been rampant and thousands of properties were built using radioactive waste in the concrete mix. Although the uranium had been extracted, the presence of radium ensured that 85 per cent of the original radioactivity remained. 4,287 properties were found to be contaminated; 5 million cu.m of material was transported to hastily designated 'official waste sites' and the total cost of the clean-up was $450 million.[23] In the old Wild West town of Durango, Colorado, 2 million cu.m of radioactive tailings covering 10 acres and a further 15 acres of ponds holding contaminated liquids left by the Vanadium Corporation of America compromised a vital tourist area. Estimates suggested a cost of over $200 million to remove and encapsulate tailings and decontaminate and protect water supplies.

In Britain, contaminated sites related to the radium and early uranium industries have been much less of a problem, although even where identified, remediation has been entirely unco-ordinated. Typically, such plants as existed were much smaller than those in the USA; however, the fact that they were smaller has probably made it more likely that, when they ceased operating, knowledge of their existence also disappeared. In some cases, the response to the discovery of information about a suspect site has been counter-productive, especially when unsophisticated dumping – largely now prohibited – was seen as the answer. This particular remedy has in the past been the worst option of all as it involved moving high-risk material from one unsuitable place to another. The government's Radioactive Waste Management Committee recognised that normal practices in dealing with materials found at Ministry of Defence sites could be accurately characterised as 'bash, burn and bury'.[24]

In 1997, the independent nuclear engineer John Large revealed that, after some years of collating a database of historical information, he had concluded that there were at least 520 sites in Britain at which radioactive materials had been manufactured or used, or where such material had been dumped. Many of these sites – some civilian, others owned by the MoD – had been involved in the production and use of luminous paint. Many had been forgotten and later sites had operated only under the guidelines of the so-called Blue Book, developed on an *ad hoc* basis by the UK Atomic Energy Authority before the regulatory system introduced in 1963 when the Radioactive Substances Act 1960 came into force. The costs of remedial action could be expected to run to hundreds of millions of pounds.[25] Most of the luminising plants (at peak, there were about 135) operated during and after the Second World War; others were established to service the motor vehicle and aircraft industries; the last such plant closed in the early 1960s. However, much of this kind of work was 'invisible'; indeed some of it was done by home-workers on a casual or subcontracted basis.[26] Identifying the sites themselves is only half the problem. Vast amounts of radioactive materials were casually dumped off-site; liquid wastes were often consigned to inadequate sumps, drains and creaking Victorian sewerage systems; and large numbers of finished products distributed for widespread use have been equally casually stored or discarded. Official practices have often been far from appropriate, as was demonstrated at the former Howards Chemical Works in Uphall Road, Ilford (built in 1899 on the site of an Iron Age fort). Amongst other items produced were luminous paint and gas mantles, using radioactive monazite sand imported from India. When the site was cleared and decontaminated at a cost of £11 million for development of the Fairfield Court housing estate in 1987, the rubble was dumped at Rainham Marshes in Essex (a kind of easily convenient 'away' place not designated as a site for radioactive waste).

Some time later, plans for a large commercial development on the Essex site were abandoned – because the site was 'found to be radioactive'. The scale of the problem indicated by John Large was made public as the government was admitting in 1997, after thirteen years of denials, that tons of radioactive waste had been secretly dumped in the Irish Sea during the 1950s and 1960s by government, universities and commercial companies. It was already admitted that a minimum of 74,000 tons of nuclear waste had been dumped in the North Atlantic (where Britain was responsible for 77 per cent of nuclear dumping). However, excluded from the official inventory was several tons of nuclear waste dumped in Beaufort's Dyke, a long-established seven-mile-long submarine trench in the Irish Sea. The critical reason for this omission was undoubtedly the fact that Beaufort's Dyke, lying below busy shipping lanes and less than six miles from the Ayrshire coast, had historically been used for the dumping of munitions and chemicals. The dump was known to contain at least one million tons of bombs, including 14,000 tons of phosgene poison-gas warheads, and Beaufort's Dyke is also close to a gas supply pipeline between Scotland and Northern Ireland.

Dumping controversies have continued, recent years. Drums of highly toxic depleted uranium (which could only have come from the huge Sellafield nuclear plant) were discovered on farmland in Northamptonshire in 1999[27] and there has been an increasing incidence of events involving the release of tritium, a radioactive isotope of hydrogen. In one case revealed in 2000, it was shown that the Ministry of Defence had colluded in the secret disposal of 'out-of-date' tritium removed from nuclear weapons (where it forms part of the triggering mechanism) on land at Shoeburyness, near Southend-on-Sea in Essex.[28] In order to avoid compliance with legislation, the 'responsible' private company that runs Aldermaston nuclear weapons plant concluded a secret disposal arrangement with the MoD, which has Crown exemption from legal dumping constraints. Particularly worrying was that land adjacent to the disposal site has since been sold for housing and farming without any public admission of 'neighbourly activity'.[29] In another shocking revelation in 1988, it was shown that British Telecom was storing tritium recovered from 2 million 'trimphones' (in which tritium activates the luminous dials) in two leaking containers installed in a public car park in Islwyn in South Wales.[30] In late 1999, the Scottish Environment Protection Agency revealed 'unexpected tritium concentrations' leaching from a dozen landfill sites, the worst examples being in Glasgow, Monklands, Aberdeen and Kirkcaldy.[31] While some of the sites were authorised for the disposal of limited low-level radioactive materials by universities and hospitals, the high concentrations were surprising; almost all of this contamination is derived from irresponsible dumping of consumer products such as trimphones and luminous 'exit' signs. It was made clear that 'concentrations ... could increase considerably in the near future.'[32]

Tritium was also involved in a different controversy in 1999 when it was alleged that a pharmaceutical plant owned by Nycomed Amersham at Whitchurch in Cardiff, which was emitting excessive tritium to the atmosphere, may have been responsible for unexplained still-births and raised infant mortality rates.[33] At the beginning of the Second World War, the Scandinavian companies Nycomed and Parmacia established a radium production operation at Amersham in Buckinghamshire. By the end of the war, about 35g of radium had been produced, along with 500kg of luminous compounds, most of it from a laboratory established in the kitchen of Chilcote House, which is now the main reception area of Nycomed Amersham plc. In 1946, the Amersham laboratory became the centre of production of radioactive materials (including tritium) for medicine, industry and research purposes and was designated the Radiochemical Centre. Three years later it came under government ownership and was operated by the UK Atomic Energy Authority; this arrangement prevailed until 1982 when the plant was the subject of the first privatisation carried out by the Thatcher government.[34]

One continuing issue involving Amersham in recent years has concerned the disputed responsibility for what used to be known as 'the national radium stock'. Although in use by hospitals and other bodies, this material was historically in the ownership of the Radiochemical Centre, but Amersham has consistently refused any responsibility for recovery, disposal or storage of these materials. In early 1989 it was suggested by a former Amersham chemist that limekiln tunnels at the Bedfordshire village of Barton-in-the-Clay were likely to be contaminated with radium.[35] Radon production had been transferred to the tunnels from the Middlesex Hospital in London as a wartime precaution in 1939, and the chemist, who had been involved in moving radioactive materials to the Ministry of Supply at the Radiochemical Centre in Amersham in 1948, claimed that considerable quantities of debris had been abandoned in the tunnels. This was a classic example of a part of the radium production industry becoming almost completely forgotten in a mere forty years, and there was considerable excitement in official circles to try to locate appropriate records, which apparently indicated that the tunnels had been returned to their original owner in a safe condition.[36] Being declared 'safe' in a frenzied abandonment during the post-war recovery period in 1948 is, however, not necessarily likely to be a reliable guide for today. Many similarly secret, and possibly unrecorded, wartime precautionary removals no doubt took place and the condition of at least one other underground radon plant, at the famous Chislehurst Caves complex in Kent, has been the subject of speculation.

Many small former military luminising plants continue to be problematical. The former naval air base HMS *Merlin* at Donibristle airfield in Fife boasted a luminising plant and many aircraft and parts were 'bashed, burned and buried' after the war. In recent years, the site has been transformed into an up-market housing development on Dalgety Bay. Unfortunately, hundreds of hotspots and radioactive particles kept turning up on the nearby beach and elsewhere, setting accusations flying and generating high levels of surveying, sampling and consultation. An assessment and report of 1996 for the former Scottish Office concluded that, while there was only a low public risk and no closed areas needed to be established, systematic long-term monitoring was essential and any physical works of any nature would need to be followed by contamination surveys.[37]

The 370-acre site of RAF Carlisle, which comprises a million square feet of warehousing, is thought to have processed 60 per cent of the RAF's radioactive waste, much of which was routinely dumped on the site. An official report to government ministers in November 2000 stated:

> Radium-226 was regarded by MoD as a classified material with no authorised disposal route. There was extensive historical use of 'bash, burn and bury' with shallow on-site disposal. Contamination, both radioactive and non-radioactive, was widespread.[38]

The worst part of this site has been described as a 'black hole' – a series of 3–4m-deep clay pits – where radioactive materials were dumped along with asbestos and other unknown contaminants. The government's advisory committee noted a general hampering of their activities by the lack of historical MoD records and pointed out that, despite the acreage and degree of the problems at the Carlisle site, the radioactive contamination was only discovered by accident in 1990.

The MoD's most radioactively contaminated site is the former Ditton Park Compass Observatory near Slough. Radium-226 was used there from 1919 until 1969, and other radionuclides such as promethium-147, polonium and tritium were probably used until the early 1970s. Approximately half the 87-hectare site was contaminated by radioactivity and substantial remedial works were carried out in the mid-1990s. About the same time, remediation was also undertaken at the former Army depot at

Stirling Forthside, where there had been two buildings housing luminising workshops. The site was contaminated by radium-226, probably by other radionuclides and by mustard gas. Serious problems were also recently discovered at the former Eaglescliffe Royal Naval Supply Depot on Teesside. Here, considerable 'bash, burn and bury' activities took place, partly on land which was subsequently sold and used for chromium production. The site was surveyed in 2000 and it was only then realised that a previous survey five years before had failed to identify the significant presence of radioactive contamination. There exist at least a dozen other similar MoD or former MoD sites which have recently been decontaminated or are still subject to remediation. Other official responses can appear unnecessarily restrictive; at least one national museum holding a collection of radium-dial instruments would not allow them to be seen by anyone, even '*bona fide* researchers'.

Difficulty also persists in dealing with civilian sites. At Wishaw in Lanarkshire, Castlehill Primary School is now a neighbour to the former site of a luminising plant operated by Smith's Industries, which produced luminous dials for the car industry. However, the local authority does not have adequate powers to serve a remediation notice on the site's owners and, while responsibility appears to rest with the Scottish Environment Protection Agency, it has claimed a lack of legally enforceable regulation. There are other civilian sites at which radium was produced, used or dumped which – far from being decontaminated – have not even been adequately researched or surveyed. Such problems are often exacerbated by lack of information and changing land ownership. The site of John MacArthur's radium plant in Runcorn has never been appropriately decontaminated; neither has the British Radium Corporation site at Limehouse; nor the site of the plant at the South Terras Mine in Cornwall. The site of the Loch Lomond Radium Works has been subjected only to partial decontamination.

References existed which rather vaguely implied the establishment by John MacArthur of a radium plant at Runcorn in Cheshire about 1910, but at the time of my own investigation in about 1989, neither the National Radiological Protection Board nor the then HM Inspectorate of Pollution had any knowledge of such premises. Quite by chance, evidence was found buried in the rate books of the parish of Runcorn; in February 1911, MacArthur had leased premises in Gas Street, Halton.[39] Gas Street extended about 125 yards from Bridge Street to the towpath of the Bridgewater Canal. The area had comprised a mixture of building types, typical of a small town: houses, small factories, a church, a public house and shops. As its name implies, Gas Street was almost wholly occupied, on both sides of the road, by the local coal-gas works – two gasholders, retort houses, purifying house, etc. The only other premises of any note, adjacent to the canal towpath and directly opposite the Victoria shipyard on the opposite bank, had most recently been used as a lead smelting works. It was these vacant premises which were taken over by MacArthur (and which he later used from the autumn of 1914 for the processing of antimony). From the 1920s, the area was occupied by a chemical factory and a tannery. Chemical production continued until the early 1970s, when the land and properties passed into the hands of the local authority prior to demolition and redevelopment.

The line of the original Gas Street now forms a boundary of a modern housing development, known as the Ellesmere Street estate, built about 1980 as part of the urban renewal of Runcorn Old Town. Given that the regulatory authorities had no knowledge of a radium plant in Runcorn, it followed that, when the site was prepared for redevelopment, appropriate surveys and decontamination did not take place. The other industrial activities in the Gas Street area would already have provided a complex toxic cocktail of contamination. Abandoned gasworks sites in a residential setting have often proved to be one of the most difficult contamination problems to handle, with such pollutants as cyanides, phenols, arsenic, lead and other heavy metals. Many by-products contain chemicals suspected of causing cancer and birth defects, and other substances are liable to spontaneous combustion and are a threat to water courses. The detritus of the earlier smelting works simply implied an additional potential source of toxic poisoning in the same area.

By 1990 the public body responsible for the redevelopment was surprisingly unable to supply relevant information; to be uncertain of long-defunct industrial activity is certainly understandable, but for it to be impossible to produce redevelopment records only a decade old seemed barely credible. A brief visit to Runcorn suggested that there was no immediately obvious radiological problem, but in the absence of information, a properly organised survey and radon monitoring seemed the appropriate means of gaining basic knowledge of the site's condition. The topography had undoubtedly changed significantly, but at the very least it would have been desirable to know what happened to the buildings and soils of the original site. If debris was recycled or dumped elsewhere, it would be proper to know where, when and under what conditions. The Contaminated Land Branch of the Department of the Environment professed no obligation or interest and referred to the responsibility of the local authority. In the further search for information, there followed protracted and difficult communication with Halton Borough Council, which seemed to react with annoyance to my findings. It was eventually established that, following my contact, the interiors of some houses had been monitored by the National Radiological Protection Board over two days in 1994, with no adverse results. Problems of this kind are of course very difficult for local authorities to assess and deal with, given the lack of records of earlier industry, and especially after redevelopment has occurred. However, it did not inspire confidence to be told by officials that the important fact to know was that 2m of contaminated soil had been removed and replaced.[40] This is precisely the kind of problem which can never be adequately addressed by a reliance on the spreading of clean topsoil.

In Scotland, the buildings and site occupied by MacArthur's Loch Lomond Radium Works remained intact after the closure of the works about 1928, apart from the removal of several tons of ores to Cornwall. Some of the buildings were used by a local joiner and builder and most of the site has been used over many years by a succession of small boat-builders. A slipway was constructed, giving boat access directly to the River Leven, which formed one boundary to the site. In 1961 discussion within the Department of Health for Scotland on the new Radioactive Substances Act 1960 provoked a 'folk memory' of the earlier survey at Balloch. This time, there was an interest in the possible hazard to health. One processing shed had been demolished and a Nissen hut erected on its concrete foundation, and other buildings were being used as stabling for ponies employed giving children pony-rides at nearby Loch Lomond-side. The new survey paid particular attention to potential problems resulting from public access. In general, radiation readings on the upper part of the site were equal to those received by radiation workers; exposure for longer than 40–50 hours per week was deemed undesirable, and a caravan-dweller on the site (for example) would receive twenty times the (then) limit for a member of the public.[41] Discussions produced a rash of memos on the nature of existing legislation; the exact powers of the Secretary of State; the problems of who had responsibility for what; and, of course, the everlasting issue of who was to pay for any eventual action. During the autumn of 1961, scientific opinion hardened, and the Radiological Protection Service of the Medical Research Council declared that the site was a radiological mess, and that ingestion, inhalation and external radiation had to be considered hazards.[42]

Inside the Department of Health, there was now an understanding that there was 'a disturbingly high level of radioactivity' and that something should be done about it. There was one memo hoping that 'there might be a Civil Defence attraction' in the use of a radioactive site, but that comprehensive clearance was favoured. In January 1963, a report appeared in the press, which a Scottish Office memo in Edinburgh noted 'with astonishment'. It was revealed that a forty-two-year-old boatman, his wife and three-year-old son were squatting in the remaining processing building. The county medical officer baulked at the situation of a young child living in a building which was well above danger level. One councillor, who lived nearby and who had been a foreman at the Radium Works, expressed astonishment that men with Geiger counters were measuring radioactivity from 300ft away when his own bedroom window was only 35ft from the yard: he would be 'very pleased to let the Medical Officer know anything he wanted.' There is no evidence that the squatting family was the subject of any radiological examination, either at the time of their eviction or subsequently.

The buildings were demolished and, in September 1963, 486cu.m of material was loose-dumped by barge off Garroch Head, in the Firth of Clyde, at a depth of 50 fathoms. It was concluded that, according to international regulations, permanent occupancy of the site could now be permitted, but maintaining the principle of the Medical Research Council of keeping exposure to a minimum, such occupancy was not recommended. All was quiet for a few more years, with boat-owners pottering about the yard, painting and repairing. Then, in April 1977, a house-building application for the adjacent Red Fox site, which had undoubtedly been contaminated by residues, provoked further activity. The authorities instructed the removed of 205cu.m of soil for disposal as low-level radioactive waste to a domestic refuse tip. Shortly afterwards, there was anecdotal evidence of bags of topsoil being offered for sale. The building proposal, for twenty-six houses and garages, went ahead without any apparent further problems.

In 1982, the owner of the site entered a planning application to build twenty-one flats on the site. The view was taken that the radiological survey which took place in 1969 was still the benchmark: there was widely distributed radioactive material on the site and the soil was thoroughly permeated with contaminated material as a result of leaching. In addition, there was now concern about the hazards of radon permeating buildings. More than ever, a comprehensive three-dimensional survey would be required, not simply one confined to surface dose-rates. The building proposal was halted and some time thereafter the ownership of the site changed hands again, although it continued to be used as a boatyard. In November 1989, during a brief personal random survey, dose-rates were found to be very variable all over the site. Readings were particularly enhanced, as they had been in 1969, near the river bank. The highest readings were found to be up to 70 microsieverts per hour in the vicinity of the slipway. Taking the then current NRPB recommended maximum dose-rate from a single site for a member of the public as 500 microsieverts per year, the maximum time limit for a member of the public near the slipway could therefore have been as low as eight hours per year. (Since that time, the recommended maximum limit for the public has been cut to 300 microsieverts per year.) The site was clearly in use by individuals for much longer than recommended times; and the site foreman used a caravan on the site as an office (*pace* the 1961 concern). Gamma spectrometry on samples of soil and vegetation taken from around the most active area suggested the desirability of further detailed radiological assessment, an occupational survey and prompt action to remedy the situation.[43] The information was passed to HM Industrial Pollution Inspectorate, the part of the Scottish Office with statutory responsibility. After some time, the inspectorate decided to take no action, indicating that the local authority was at liberty to act on its own account.

A comprehensive survey was begun jointly by Dumbarton District Council and Strathclyde Regional Council in April 1991 and reported five months later. The survey comprised the measurement of surface contamination and dose-rate measurements taken 1m above ground. A 2m grid pattern was established for the main site and part of the adjacent hotel grounds. Spot dose-rates and radon measurements were also taken on the Red Fox housing development and in two other older houses at the entrance to the main site, Bankhead and Tighnairn. A number of measurements at various points elsewhere in Balloch were made for comparison purposes. The whole site was found to be contaminated with radium-226 and its daughters and:

> the surface dose-rates are considerably greater than the 2–5 μSv/hour measured in the same area in 1969 by a National Radiological Protection Board Survey team. The cause of this increase in dose rate at the ground surface appears to be erosion of the ash cover material put down after the last clearance operation.[44]

It was stated that, 'as dose rates vary by an order of magnitude across the site, and because many types of occupation could be envisaged, a range of potential doses to people using the site are possible'.[45] Radon levels at the Red Fox development were below the official action levels, but appeared to breach regulations in Tighnairn. Consideration was given to the commissioning of a sophisticated three-dimensional radiometric survey, but the cost of this work alone was much too great for the local authority, without even contemplating any subsequent remedial work. The slipway area was stabilised and concreted, and a further survey in 1997 revealed a significant reduction in dose rate; the present West Dunbartonshire Council continues to monitor the site five times a year in order to confirm its integrity. Planning constraints have been applied to the main site and development on neighbouring sites discouraged.

In considering exposure to radiation contamination, the National Radiological Protection Board recognises that we should 'assume a progressive increase in risk with increasing dose, with no safe threshold'. Previous notions of 'efficacious' radiation levels, which gave way to 'safe', then 'tolerable', are at last being abandoned. In the autumn of 1997, new evidence was published which suggested that previously unknown pathways could be leading to much greater low-level radiation damage than had previously been suspected.[46] Other countries are much more interventionist than Britain in dealing with radiation-contaminated land. The USA has been spending vast sums on decontamination in recent years, although new sites are still being rediscovered. In France, the National Agency for the Management of Radioactive Waste produces an annual inventory of early industrial sites. In 1997 it contained information on over 1,100 sites with a contamination equivalent to 20mg of radium. This proactive agency mounted a public campaign to recover the thousands of watches and clocks still in circulation with radium-luminous dials. However, the inventory was hampered by the fact that sites were only listed if owners volunteered relevant information. There was one well-known site at Saint-Denis in the north of Paris which was unlisted despite the fact that it had been a radium extraction plant; workers in the existing butchers' waste factory parked their cars in the basement, where there was known to be a significant amount of buried radium waste.

Proposals by the UK government to establish registers of contaminated land were scrapped following pressure by vested interests such as developers and financial institutions. The original intention was hardly ideal, but was at least a step in the right direction. Opposition was based on the assumption that identified land would be forever blighted, since inclusion on the proposed register would continue even after remedial work. The idea that 'the polluter pays' has proved to be little more than a political slogan which has little prospect of practical success. The House of Commons committee which reported in 1990 on contaminated land was disappointed by the lack of information on the subject offered by the Department of the Environment. The first sentence of the report ominously began, 'We do not wish to be unnecessarily alarmist...'. One of the committee's main conclusions at least points to a better approach:

> Pollution of the environment as a consequence of industrial activity is a problem which must receive a positive response. It is not satisfactory to define such pollution out of existence by claiming that it arises only in a narrow set of circumstances; nor should we claim that a site has been cleaned up when it has merely been covered with a layer of clean topsoil. [47]

The questions of what to do with radioactive waste, and where, are as far from solution as ever. In the USA, the Department of Energy has been studying the potential for a huge national underground repository at Yucca Mountain, 100 miles north-west of Las Vegas. Already, many billions of dollars have been spent and there is no prospect of anything being stored there until 2010 at the earliest. In the meantime, there are substantial protest movements against the proposal and continuing geological and scientific objections. It is entirely possible that the Yucca Mountain scheme will have cost a fortune in gestation, but will fail to reach maturity. The issue appears no nearer a solution in any other country and large volumes of highly radioactive materials are building up at a huge variety of hopelessly inadequate sites. A politically inspired British plan for a similar, but much smaller, underground 'rock laboratory' near the huge nuclear site at Sellafield in Cumbria did not long survive scientific objections. At the same time, the low-level waste site at nearby Drigg is reaching the stage at which it will no longer be able to accommodate further material.

The industry is not only secretive and indecisive; it is dangerously deceitful. At Dounreay, on the coast of northern Scotland where a fast-breeder reactor was sited, unknown types and quantities of radioactive material were routinely dumped into two shafts over many years. In 1977, one exploded, scattering plutonium and other unknown substances widely in the environment; the regulatory authorities were not told of this incident for eighteen years. In the summer of 2004 it was revealed that, contrary to stated policy, the government was secretly storing 10,000cu.m of high and intermediate-level nuclear waste belonging to foreign countries which it could no longer process and which was 'too expensive' to return. Meanwhile, Britain's own inventory of 10,000cu.m of high-level waste and 250,000 tonnes of intermediate waste was being stored in deteriorating buildings at Sellafield and at least nineteen other sites around the country.[48] If this is normal behaviour in a responsible, democratic Western country, what is to be expected elsewhere, and why should the industry and its political masters be trusted?

In the face of these problems, some desperate measures are apparently being contemplated by the government committee appointed to examine the crisis of the UK's radioactive waste disposal. It is disturbing, to say the least, that some of the methods being considered adopt such outdated, irresponsible and dangerous 'throw-it-away' concepts as burial at sea; allowing the wastes to melt their way though Antarctic ice-sheets; placing the waste where active tectonic plates will grind it into the Earth's mantle; or, in true science-fiction comic style, firing the wastes into space, either to orbit outside the solar system or to be consumed by collision with the sun.[49]

In parallel with these deliberations, and in a climate of some desperation over energy supplies, the UK government seems likely to restart a new era of reliance on nuclear power generation, which has been allowed to run down in recent decades for practical and financial reasons. In common with other equally serious issues, it seems depressingly the case that decision-making is avoided and delayed until hard reality demands that courses of action are finally determined in a climate of desperation. As the Indian prime minister, P.V. Narasimha Rao, said in an early interview after assuming office in 1991, 'Decisions are easier, you know, when there are no choices left.'[50]

EPILOGUE

Radium might have seemed to be a dead subject, long forgotten and of no real consequence, but this book may have helped, if not to change that impression, at least to throw light again on the subject. Its discovery and investigation by the Curies were undoubtedly examples of scientific excitement and endeavour of an unusual order; and who, having read even such a brief account as this one, could forget the story of the women dial-painters of New Jersey and Illinois? Their struggle, and the momentous social changes that were won, against all the political and financial odds, are surely important examples of benefit to the common good arising out of the misery of individuals. As the twentieth century moved on, the relatively new sciences of industrial medicine and the prevention and alleviation of occupational disease were given an impetus that ensured public attention and support. Who knows how medicine – and in particular, the treatment of that appalling scourge, cancer – would have progressed without the avenues that were opened up by the use of radium? It is an extraordinary feature of that disease that it still wreaks a terrible toll after such a long history of collaborative international effort against its worst effects, but the contribution made by radium treatments and the more sophisticated radiotherapies that followed has been incalculable. It seems likely that the progress brought during the first decades of the twentieth century by the use of radium will be surpassed in the first decades of the twenty-first century by equally exciting developments in genetic medicine.

The conundrum hinted at by radium – 'which rivals the sun in giving out everlasting heat and brilliant light' as a popular science magazine claimed in 1900[1] – was the hint of the transmutation of elements. When it became known that the processes of radioactivity released more energy than any other known phenomenon, it led to luminaries as celebrated and careful as Frederick Soddy declaring in his seminal book, *The Interpretation of Radium* that, 'a race which could transmute matter would have little need to earn its bread by the sweat of its brow'.[2] From the beginning of the radium story, such enthusiasm was picked up by others and exaggerated beyond reason. At one point, the science editor of the *New York Times* promoted the idea that a single building the size of a small-town post office would produce the entire energy requirements of the USA. Not only that, but there would be by-products such as gold, which would be so cheap and plentiful that it would be used as a roofing material.[3] Similar misplaced passion, which continued with the much later, widely believed and deceitful British slogan 'electricity too cheap to meter', has dogged the entire history of the nuclear industry ever since. The fact that radium was enormously expensive was not only always faithfully reported, but it somehow seemed to convey an important feature of the mystique; the idea that its value was poetically (but also almost literally) 'beyond compare' apparently conferred enigmatic additional benefit.

Even the simpler of the problems associated with the use of radium have continued to resurface. Perhaps the commonest form of radium available to the public was radium-water, and we have seen what some of the implications of that were. Yet still people are ingesting radium from bottled water, albeit without the excited deliberation of those impressionable times. Recent studies in Hungary have concluded that in three out of eighteen commercial bottled waters the radium content breached World Health Organisation recommendations.[4] Concern over this issue led the Food Standards Agency in the UK to conduct studies of the radiation content of 175 samples of natural mineral water, spring water and bottled drinking water. Samples were tested for various isotopes of uranium, polonium, thorium, tritium, radium and lead; although no breaches of legal limits were found, nor evidence requiring dietary intervention, the matter is nevertheless one which will require continued monitoring at an international level.[5]

As the world has moved from the radium age into a more complex nuclear age, people have become more sceptical, if not cynical, of the realities involved. Even in the relatively prosaic business of electricity generation, the technology is difficult and the fully audited costs probably prohibitive. The vital problems of decommissioning, waste disposal and the potential for misuse of nuclear materials are still probably as far from being solved as ever. There is an increasing lack of trust in political decision-making, and a perception that public interest is too often compromised by commercial imperatives. This problem manifests itself most noticeably in relation to science and technology – the nuclear industry and the handling of issues such as BSE, MMR, nanotechnology and genetic modification are typical current concerns over which the public has been increasingly unwilling to accept what 'official' sources say at face value. 'Alternative' sources of information and opinion typically grow exponentially, with the detrimental end result that opinions become more and more polarised.

The climate is such that in late 2004 there was increasing discussion of the desirability of establishing some form of 'risk commission', independent of government, which would analyse such issues and, by providing independently researched information, attempt to inspire agreement and confidence and would counter the negative effects of the polarisation of attitudes. With consideration being given to a new phase of nuclear energy provision, hard thinking and serious decision-making are now necessary. It is inconceivable that society should continue to exploit nuclear technology in the politically and economically unaccountable way that has hitherto been permitted. If a broad consensus cannot be achieved by which vital decisions are made, further decades of incompetent piecemeal national energy policy will result. The ever-insightful Parkinson's Law illustrates what happens (and what has already happened) when real issues are ducked or fudged:

> The man who is denied the opportunity of taking decisions of importance begins to regard as important the decisions he is allowed to take. He becomes fussy about filing; keen on seeing that pencils are sharpened, eager to ensure that the windows are open (or shut) and apt to use two or three different-coloured inks.[6]

It is a paradox of our times, and of science, that pitchblende – the mysterious substance which had for decades been seen as a troublesome, useless waste material – brought us all into uncomfortable association with the complexities of a nuclear age whose international future has been jeopardised because of its inability to handle its own deadly waste.

APPENDIX I

COST AND PRODUCTION OF RADIUM

World market price for radium between 1904 and 1930 in US dollars per milligram: *6

1904	10–15
1905	25–50
1906	60
1909-10	75–135
1911-12	150
1912-14	180
1915	160
1916-1922	120–150
1923-30	70

Total amount of radium produced by country of origin, 1898–1928, in grams.

United States	250
Belgian Congo	245
Czechoslovakia (Austria)	45
Portugal	15
Madagascar	8
Russia	6
UK (Cornwall)	5
South Australia	1
Total	575g

Figures for both tables from the Uranium Institute, London

APPENDIX II

THE URANIUM-238 DECAY CHAIN

Symbol	*Element*	*Radiation*	*Half-Life*	*Decay Product*
^{238}U	uranium-238	alpha, gamma	4.5 billion years	^{234}Th
^{234}Th	thorium-234	beta, gamma	24.1 days	^{234}Pa
^{234}Pa	protactinium-234	beta, gamma	1.17 minutes	^{234}U
^{234}U	uranium-234	alpha, gamma	247,000 years	^{230}Th
^{230}Th	thorium-230	alpha, gamma	80,000 years	^{226}Ra
^{226}Ra	radium-226	alpha, gamma	1,602 years	^{222}Rn
^{222}Rn	radon-222	alpha	3.82 days	^{218}Po
^{218}Po	polonium-218	alpha, beta	3.05 minutes	^{214}Pb
^{214}Pb	lead-214	beta, gamma	27 minutes	^{214}Bi
^{214}Bi	bismuth-214	alpha, beta, gamma	19.7 minutes	^{214}Po
^{214}Po	polonium-214	alpha	0.00016 second	^{210}Pb
^{210}Pb	lead-210	beta, gamma	22.3 years	^{210}Bi
^{210}Bi	bismuth-210	alpha, beta	5.01 days	^{210}Po
^{210}Po	polonium-210	alpha, gamma	138.4 days	^{206}Pb
^{206}Pb	lead-206	none	stable	none

APPENDIX III

THE PROPERTIES OF RADIUM

The heaviest of the alkaline earth elements, radium is over a million times more radioactive than the same mass of uranium. The metal is brilliant white in colour but blackens on exposure to air; it exhibits a faint blue luminescence, as do its salts; it decomposes in water and in combustion colours a flame red. The elemental metal is relatively rare, and the word 'radium' normally refers to the compounds radium chloride ($RaCl_2$) or radium bromide ($RaBr_2$), both crystalline solids which resemble common salt.

Radium is a decay product of uranium and is therefore to be found in minute quantities in all uranium-bearing ores, principally pitchblende (an impure uranium oxide) and carnotite (a mineral containing uranium, vanadium and lead). The proportion of radium found in the richest ore amounts to only one part in several million. Radium is also present (in amounts incapable of extraction) in mineral springs, rocks, soil, sea-water and air.

Radium has the chemical symbol Ra, the atomic number 88 and the atomic weight 226.0254. Its melting point is 973K (1,292°F) and boiling point 2,010K (3,159°F). Radium emits three kinds of radiation – alpha, beta and gamma; its most stable isotope, radium-226, has a half-life of 1,602 years and decays to the radioactive gas radon, and eventually to inactive lead. There are in total twenty-five isotopes, of which four (^{223}Ra, ^{224}Ra, ^{226}Ra and ^{228}Ra) are found in nature as decay products of uranium or thorium.

Radium, and all its preparations, is notable for being exothermic: it gives off heat. A gram of radium gives off more than enough heat in one hour than is required to raise its own weight of water from freezing to boiling point; this is almost a thousand times the energy produced by the explosion of a gram of nitroglycerine. Despite this spontaneous production of energy, the radium remains effectively unchanged.

In use, radium was extremely expensive due to the rarity of exploitable ores and the complexity of the chemical processing. There were perceived beneficial medical effects in the use of radium (usually in the form of the daughter product, radon) associated with the effects of energetic heavy particles, similar to X-rays. However, lack of understanding and misuse masked for many years the fact that the effects of radium on the human body could be debilitating and lethal, leading to infections, bone necrosis, tumours and carcinomas in various organs, anaemia and other biological alteration. Radium in all its forms was then recognised as a hazard to the human body, whether acquired by inhalation, injection or by exposure from external sources.

APPENDIX IV

PLACES IN BRITAIN ASSOCIATED WITH RADIUM

This appendix lists those places having a particular connection with radium. There were very few locations at which radium ores were mined, or where radium was chemically separated, and such places are all included; there would be little point in listing all the places where radium was used (hospitals, university laboratories, military depots, etc), but sites with a more unusual connection or that are, or have been, liable to particular contamination problems are noted.

MINING

Uranium minerals were found mostly in Cornwall, and prior to the discovery of radioactivity and of radium, were initially used as glass colouring agents.

Buckfastleigh Estate, Ashburton, Devon
Extremely rich pitchblende deposits identified but not exploited (site regarded as dangerous)

Camborne, Cornwall
Torbernite found at Tincroft Mine, *c.*1805

St Ives, Cornwall
Pitchblende found at Wheal Trenwith, *c.*1843

St Just, Cornwall
Pitchblende identified at Wheal Edward and Wheal Orrles, but neith erexploited

St Stephens/Grampound, near St Austell, Cornwal
Torbernite found at South Terras Mine, 1873: this proved the most productive British source of uraniumand radium

PROCESSING

Balloch, Dunbartonshire
Location of the Loch Lomond Radium Works, operated by John S. MacArthur from 1915 to 1922. The site has not been fully decontaminated and has occasionally been the focus of concern

Halton (Runcorn), Cheshire
Location in Gas Street of Radium Works of John S. MacArthur, from 1911 to 1915. This was a 'lost' site, which was never decontaminated before its redevelopment for housing

London
Location of British Radium Corporation radium works, at Thomas Street, Limehouse: this was never appropriately decontaminated before redevelopment. Later works were built, but possibly unused, at Croydon Road, Elmer's End

St Austell, Cornwall
Location in Trevarthian Road of a small processing plant briefly operating *c.*1928

St Stephens/Grampound, near St Austell, Cornwall
A small processing plant was built at South Terras Mine operating sporadically *c.*1922–31

MISCELLANEOUS

Aldermaston, Berkshire
Nuclear weapons plant where secret radiation experiments on humans took place from the mid-1950s until the late 1980s

Amersham, Buckinghamshire
Site of the National Radiochemical Centre, later the nuclear medicine group Amersham International, now known as Nycomed Amersham plc

Barton-in-the-Clay, Bedfordshire
Location of secret wartime radon production facility for medical purposes
Bath, Somerset
Location of helium experiments in the Hot Springs, by Dewar, Strutt & Thomson: King's Well developed by Ramsay as a 'Radium Inhalatorium'

Beaufort's Dyke
Submarine dumping ground in the Irish Sea, which contains radioactive waste, munitions and chemicals

Buxton, Derbyshire
Location of helium experiments by Dewar & Strutt: the Hydropathic Hotel was the office of Radium Ltd, which sold radium-water treatments, some imported from Germany

Cardiff (Whitchurch), Glamorgan
Site of Nycomed pharmaceuticals plant allegedly venting radioactive tritium to the atmosphere

Carlisle, Cumbria
The huge RAF base here has suffered extensive radiation contamination

Carrickfergus, Northern Ireland
Location of plant producing radioactive fertilisers, salt, bath soap, antibacterial agents and photographic materials

Dalgety Bay, Fife
The site of the former Donibristle airfield luminising plant; the surrounding public area has experienced continuing radium contamination

Drigg, Cumbria
Site of the largest official UK radioactive waste disposal site, near Sellafield; now nearing capacity. Many other landfill sites throughout the country, some official, others totally unregulated, have been used for the dumping of radioactive wastes: see Chapter 11

Eaglescliffe, Teesside
Site of former RN depot has been badly contaminated with radiation and heavy metals

Harwell, Oxfordshire
Government nuclear laboratory where secret radiation experiments on humans took place from the mid-1950s until the late 1980s

Haywards Heath, Sussex
Location of Radium Electric Ltd, producing the Q-Ray Electro-Radioactive Compress and related products

Ilford, Essex
Site of radiation-contaminated factory producing gas mantles (see also Rainham Marshes)

London
Location of the Radium Institute, in Ridinghouse Street, off Portland Place
Location at 62 Oxford Street of Radium Treatments Ltd
Location at Edmonton, north London, of plant producing Sparklets Radon Bulbs and other radioactive products

Manchester
The Christie Hospital is the location of the important Holt Radium Institute

Rainham Marshes, Essex
Large dumping area, used for radium-contaminated wastes, heavy metals and other pollutants

Slough, Berkshire
The former Ditton Park Compass Observatory has been the most severely radium-contaminated Ministry of Defence site

Stirling, Stirlingshire
The Forthside army luminising depot has been badly contaminated with radium and heavy metals

Wishaw, Lanarkshire
Adjacent to a primary school, the former Smiths Industries luminising plant is heavily radium-contaminated

THE RADON PROBLEM

The problem of the naturally occurring radioactive gas radon is a continuing concern, and current estimates are that this is responsible for 2,000–3,000 cancer deaths per year in the UK. Certain parts of the country (Devon, Cornwall, Somerset, Derbyshire, Northamptonshire and Highland and Grampian Regions in Scotland) are especially susceptible, due to geological conditions. Strict new building regulations take account of this issue, and national and local government are concerned to alleviate the problem, essentially by the use of a variety of under-floor ventilation procedures.

APPENDIX V

RADIUM AND HUMANS

Despite its mystique and its monstrous popular image, radiation is a fundamental process of nature that has been in action since the birth of the universe and was not in any way changed by its mere discovery a little over a century ago. *Non-ionising radiation* (which does not concern us here) includes light, heat, radar, radio waves and microwaves. The type of radiation produced by radium is *ionising radiation*, which is the result of the process by which unstable isotopes (or forms of individual elements) transform themselves by the exchange of their subatomic electrons into more stable forms. Humans are constantly bombarded by natural and man-made ionising radiation from hundreds of sources, and radioactive disintegration is continually occurring inside the body; however, a vital feature of this process for humans is that it is not detectable by any of our senses.

In simple terms, 'radioactive decay' produces three different forms of radiation, and they are the constituent radiations of radium:

Alpha rays are nuclei of helium; the particles are heavy, positively charged and 'slow-moving' (even if they do move at 15 million metres per second); they cannot travel great distances and they lose energy rapidly; they can be stopped by a sheet of paper and are therefore very easily shielded.

Beta rays are positively charged electrons which travel at about 60 per cent of the speed of light; they are more penetrating than alpha rays, passing though paper or metal foil without loss of energy, but are relatively easily stopped by lead shielding.

Gamma rays have no mass or charge but are radiations similar to light and almost identical to X-rays; they have very short wavelengths and very high energy; they are deeply penetrating and can be shielded only with difficulty.

These three forms of radiation can damage the human body in different ways as they transfer energy in passing through tissue; this can result in abnormal chemical reactions or molecular and genetic changes. A substance emitting alpha rays can safely be held in the hand, as the alpha radiation is stopped by the outer layer of skin; beta rays are capable of penetrating several millimetres of tissue, so are potentially dangerous to superficial tissue but not to internal organs; gamma rays, having high energy and the ability to penetrate deeply, can pass completely through the human body and are therefore of great hazard in every instance to external and internal organs. Despite the great penetrating ability of gamma radiation, possibly after brief exposure by an external source, this characteristic does not imply that gamma radiation is necessarily the most dangerous to humans; the ingestion of particles emitting alpha or beta rays inside the body is likely to be extremely dangerous as such particles can remain lodged in tissue and continue to irradiate internal organs indefinitely.

There are complex relationships between the various units used to express radiation values and biological effects. In cases of the ingestion of radioactive particles, where direct measurement is impossible, *absorbed dose* is calculated on the basis of the mass of tissue involved. Unfortunately, equal absorbed doses do not necessarily have equal biological effects, partly due to the differences in energy and charge of different radiations, and because the body's reaction is dependent on the nature of the tissue irradiated. A weighting factor, based on the harmfulness of different radiations is applied to give the *dose equivalent.* (For example, the factor for the seemingly most dangerous gamma radiation is 1, whereas the factor for actually more dangerous alpha particles is 20). A further factor can be applied which takes into account the expected biological effect on different body organs – this gives the *effective dose equivalent*, which would give the same risk of genetic damage if applied uniformly to the whole body. This calculation can be further modified to give the *collective effective dose equivalent* which is used referring to groups of workers or larger populations.

The absorbed dose is measured, per unit mass of tissue, in a unit named the gray (Gy). The additional factors applied as described above produce a unit named the sievert (Sv) or millisievert (mSv). For example, the factor for alpha radiation is 20, so an absorbed dose of 1Gy from alpha radiation becomes a dose equivalent of 20Sv. Further factors applied as described continue to produce a figure expressed in sieverts or millisieverts; these measurements allow a wide variety of circumstances to be represented as a single number usually accepted as being a sufficiently accurate indicator of health risk from different radiations in differing situations.

NOTES AND REFERENCES

CHAPTER ONE

1 'Le gisement d'urane de St-Joachimsthal' by Paul Gaubert, in *Le Radium*, January 1906, pp.1–9

2 It is thought that use of the word 'cobalt' in reference to a mineral originated in Saxony, and derived from the German word *Kobelts* meaning gnomes or goblins

3 'Radiological aspects of former mining activities in the Saxon Erzgebirge, Germany' by Gert Keller, in *Environment International*, vol.19, 1993, pp.449–54

4 *De re metallica* by Georg Bauer (Agricola), 1912 translation, p.214

5 *De re metallica* by Georg Bauer (Agricola), 1912 translation, p.214

6 *De re metallica* by Georg Bauer (Agricola), 1912 translation, p.217

7 'A history of pitchblende' by Dr J.R. Morgan, in *Atom* No.329, March 1984: United Kingdom Atomic Energy Authority

8 'The menace under the floorboards' by A.F. Gardner, R.S. Gillett and P.S. Phillips in *Chemistry in Britain*, London, April 1992; and correspondence with Dr James Stebbings

9 'Radiological aspects of former mining activities in the Saxon Erzgebirge, Germany' by Gert Keller, in *Environment International*, vol.19, 1993, pp.449–54

10 'Uranium's scientific history, 1789–1939'. Paper by Dr Bertrand Goldschmidt, Uranium Institute, London, Sept. 1989 (see www.uilondon.org/ushist.html)

11 'Radium minerals of Saxony and their discovery' by Prof. C. Schiffner, in *Mining and Engineering World*, Chicago, 15 June 1912

12 'A history of pitchblende' by Dr J.R. Morgan, in *Atom* no.329, March 1984: United Kingdom Atomic Energy Authority; and 'Uranium's scientific history, 1789–1939'. Paper by Dr Bertrand Goldschmidt, Uranium Institute, London, Sept. 1989

13 'Radium' by Richard B. Moore, US Bureau of Mines, in *Transactions*, AIMME, vol.60, 1919; also 'Radium dreams' by E. de Hautpick, in *The Mining Journal*, London, February 1913

14 'Radium' by Richard B. Moore, US Bureau of Mines, in *Transactions*, AIMME, vol.60, 1919; also 'Radium dreams' by E. de Hautpick, in *The Mining Journal*, London, February 1913; and 'Metallurgy of radium and uranium' by H.A. Doerner, in *Handbook of Non-Ferrous Metallurgy*, McGraw-Hill, New York, 1945
15 'Teeth made of stone', in *Scientific American* vol.2 no.51, 11 September 1847
16 *Scientific American*, vol.12 no.12, 18 March 1865
17 'A history of pitchblende' by Dr J.R. Morgan, in *Atom* no.329, March 1984: United Kingdom Atomic Energy Authority
18 Database of mines in Devon and Cornwall by Heather Coleman (1999): www.dawnmist.demon.co.uk/minedata.htm
19 'Uranium and radium in South West England' by Neil Dickinson, in *Journal of the Plymouth Mineral and Mining Club*, vol.11 no.2: September 1980
20 'The South Terras radium deposit, Cornwall' by T. Robertson and H. Dines in *The Mining Magazine*, London, September 1929
21 'Uranium and radium in South West England' by Neil Dickinson, in *Journal of the Plymouth Mineral and Mining Club*, vol.11 no.2: September 1980
22 'Uranium and radium in South West England' by Neil Dickinson, in *Journal of the Plymouth Mineral and Mining Club*, vol.11 no.2: September 1980
23 Presidential address by C.V. Smale, *Journal of the Royal Institution of Cornwall*, vol.1, 1993

CHAPTER TWO

1 General biographical information on Marie Curie from numerous published sources, including *Madame Curie* by Eve Curie, da Capo Press, Doubleday, Doran & Co., New York, 1904 and 1937; also *Grand Obsession: Marie Curie and her World* by Rosalynd Pflaum, Doubleday, New York, 1989; *Marie Curie, a Life* by Susan Quinn, Heinemann, London, 1995; and *Marie Curie and the Science of Radioactivity* by Naomi Pasachoff, Oxford University Press, 1996
2 Marie Curie, *Pierre Curie* and autobiographical notes, Macmillan, New York, 1923, p.161
3 *Marie Curie, a Life,* by Susan Quinn, p.65
4 Marie Curie, *Pierre Curie* and autobiographical notes, Macmillan, New York, 1923, p.74
5 'The light and brilliancy of Marie Curie' by Peter Craig, *New Scientist*, 26 July 1984, p.33
6 Quin, *op. cit.* p.140
7 William Crookes, *Proceedings of the Royal Society*, 83, xx, 1910
8 Henri Becquerel, 'Emission of new radiations by metallic uranium' in *Chemical News*, London, 26 June 1896, p.295
9 Quin, *op. cit.* p.144
10 P. Curie and Mme P. Curie, *Comptes rendus de l'Académie des Sciences*, Paris, vol.127, p.175
11 *Madame Curie* by Eve Curie, p.167
12 *Ibid.* p.168

13 From Marie Curie's doctoral thesis of 1903, published as *Recherches sur les substances radioactives*' (*Radioactive Substances*, Dover edition, p.20–21)
14 'Radium and radioactivity' by Marie Curie, *Century Magazine*, January 1904, pp.461–66
15 Marie Curie in a letter, quoted in *Madame Curie* by Eve Curie, p.169
16 Quin, *op. cit.* p.153
17 Marie Curie, *Pierre Curie*, *op. cit.* p.103
18 Quin, *op. cit.* p.157
19 From a lecture by N. Fröman to the Swedish Academy of Sciences, Stockholm, in February 1996, published by the Nobel Foundation
20 *Grand Obsession: Marie Curie and her World* by Rosalynd Pflaum, p.94
21 'Radium and radioactivity' by Marie Curie, *Century Magazine*, January 1904, pp.461–66
22 *The Times*, 25 March 1903, p.10
23 Pflaum, *op. cit.* p.103
24 Frederick Soddy, 'Method of applying the rays from radium and thorium to the treatment of consumption' in *Nature*, no.1761, vol.68, 30 July 1903, p.306
25 *Nature*, vol.69, p.5
26 *Nature*, vol.68, pp.90 and 343. Lord Blythswood (Sir Archibald Douglas Campbell, 1835–1908) was a close friend of Lord Kelvin and an eminent 'gentleman-physicist' with a well-appointed laboratory at his home beside the River Clyde

CHAPTER THREE

1 Pierre Curie's comment to William J. Hammer in *Radium and other radioactive substances*, by William J. Hammer, 1904, p.27
2 *Marie Curie, a Life*, by Susan Quinn, p.206
3 Lecture by Prof. N. Fröman, Royal Swedish Academy of Sciences, Stockholm, February 1996, published by the Nobel Foundation
4 Account from Marie Curie's doctoral thesis of 1903, published as *Recherches sur les substances radioactives* (*Radioactive Substances*, Dover edition, p.67)
5 Pearce's account to the Colorado Scientific Society in 1895, quoted in 'Buried treasure to buried waste', by Edward R. Landa, *Colorado School of Mines Quarterly* vol.82 no.2, Summer 1987, p.5. (Pearce was back in England by that time, but having brought 'radium crystals' with him from the natural springs at the mining town of Empire, Colorado, he added the supposedly therapeutically radioactive material to everything he drank until his death at the age of ninety in 1927.)
6 'Buried treasure to buried waste', Landa, *Colorado School of Mines Quarterly* vol.82 no.2, Summer 1987, pp.9–10. See also 'Early uranium mining in the United States', a paper by F.J. Hahne for the Uranium Institute in London, September 1989: www.world-nuclear.org/usumin.htm. Uranium and vanadium are still widely used as alloy additives, with chromium, nickel, manganese and other elements, in the production of special steels for military and aerospace purposes. Other uses are in petroleum cracking, organic chemistry, ceramics and textile dyeing
7 *Radioactivity in America*, by Lawrence Badash, p.137

8 US Geological Survey 1917, p.833 (quoted in 'Buried treasure to buried waste', Landa, *Colorado School of Mines Quarterly* vol.82 no.2, summer 1987, p.16). Radium in the metallic state was first isolated by Marie Curie in 1910
9 *Nature*, 13 August 1903, p.343. J.J. Thomson, Cavendish Professor of Physics at Cambridge, is generally credited as being the discoverer of the association of radioactivity and natural waters
10 *Nature*, 28 May 1903, pp.90–91
11 R.J.Strutt, letter of 30 December 1903, in *The Discovery of Helium and Radium in the Hot Springs of Bath, England*, by Prof. Sir James Dewar FRS, and Hon. R.J. Strutt (undated)
12 *Nature*, 16 July 1903
13 Letter from Alexander Graham Bell to Dr Z.T. Sowers of Washington, published in *American Medicine*, August 1903
14 Marie Curie, quoted in *Madame Curie*, by Eve Curie, p.204
15 After Ramsay's unique discovery of the five inert gases, argon, helium, neon, krypton and xenon, he had collaborated with Soddy to determine that helium was produced by radium, thus confirming Rutherford's theory of disintegration
16 *Engineering*, London, November 1907
17 Letter from anonymous scientist, *The Times*, London, 19 September 1908
18 *Radioactivity in the Environment*, by Ronald L. Kathren, p.13
19 *Punch* magazine 1904, quoted in *The Infancy of Atomic Physics – Hercules in his cradle*, by Alexander Keller, p.107
20 *The Times*, 27 October 1908, p.13
21 *Nature*, London, 17 February 1910
22 *Nature*, London, 17 February 1910
23 *The Mining Journal*, London, 8 May 1909
24 'Radioactive minerals in Russia' by E. de Hautpick, in *The Mining Journal*, 25 February 1911, p.185–7
25 *The Times*, London, 4 December 1912
26 *Glasgow Herald*, 9 February 1909
27 *The Lancet*, London, 12 June 1909
28 *Nature*, London, 17 February 1910
29 The Limehouse Cut is a 2-mile long canal built in the mid-eighteenth century, linking the River Lee at Bromley-by-Bow with the River Thames at Limehouse; it avoided a long loop in the Thames around the Isle of Dogs
30 *The Life of Sir William Ramsay KCB, FRS*, by Morris W. Travers, 1956, p.227
31 Reported in *The Mining Journal*, London, 23 October 1909
32 *Glasgow Herald*, 27 January 1910
33 *The Mining Journal*, London, 29 January 1910
34 *The Lancet*, London, 5 March 1910, p.685
35 Files of Radium Ltd, London, PRO: BT31/19554/110612
36 G.B. Shaw, from *Preface on Doctors*, a long introduction to his play *The Doctor's Dilemma*, first published in 1906; reproduced by permission of the Society of Authors and the Bernard Shaw Estate

37 *The Times*, 20 October 1910
38 J.S. MacArthur, 'The extraction of radium' in *The Mining Magazine*, London, February 1916, p.86
39 J.S. MacArthur, 'The extraction of radium' in *The Mining Magazine*, London, February 1916, p.87
40 Files of the International Vanadium Co. Ltd. PRO: BT31/18883/103460
41 *The Times*, 8 January 1914
42 'Radium dreams', by E. de Hautpick, in *The Mining Journal*, London, 8 February 1913, p.134
43 'Radium dreams', by E. de Hautpick, in *The Mining Journal*, London, 8 February 1913, p.135
44 Flannery, quoted in *The Radium Girls*, by Claudia Clark, p.52
45 Field, from 'The efficiency of radioactive waters for the control of faulty elimination', in *Medical Record* 87 (1915) p.390–94; quoted in *The Radium Girls*, by Claudia Clark, p.52
46 Field, in *American Medicine*, January 1926, pp.40–43.
47 'Buried treasure to buried waste', Landa, *Colorado School of Mines Quarterly* vol.82 no.2, summer 1987, p.36
48 Dr C.G. Davis in *The American Journal of Clinical Medicine*, quoted in *Radioactive Curative Devices and Spas* by Paul W. Frame, Oak Ridge Associated Universities, 1989

CHAPTER FOUR

1 J.S. MacArthur, 'The radium industry and reconstruction' in *The Mining Journal*, London, 18 January, 1919
2 J.S. MacArthur, 'The extraction of radium', in *The Mining Magazine*, London, February 1916
3 Swansea Radium Ltd. PRO: BT31/22506/137602
4 Meeting at the Merchant's House, Glasgow, reported in the *Glasgow Herald*, 7 April 1914
5 Meeting at the Merchant's House, Glasgow, reported in the *Glasgow Herald*, 7 April 1914
6 *Glasgow Medical Journal*, May 1914
7 *Glasgow Medical Journal*, May 1914
8 Minute Book, Glasgow and West of Scotland Radium Committee: Glasgow University Archives
9 Minute Book, Glasgow and West of Scotland Radium Committee: Glasgow University Archives
10 Minute Book, Glasgow and West of Scotland Radium Committee: Glasgow University Archives
11 The mediaeval village of Gif is now a city with a very large scientific campus, which still includes centres for nuclear studies; the original radium plant was closed in 1957, and its site is still contaminated. The director of the 'Laboratoire d'essais des substances radioactives' was Jacques Danne, who had been one of Marie Curie's assistants, and who was also editor-in-chief of the journal *Le Radium*

12 *The Cornish Guardian*, 15 January 1904
13 See 'Radium produced in France from Cornish ore' in *The Mining Journal*, London, 22 November 1913; 'The South Terras radium deposit, Cornwall', by Robertson and Dines, in *The Mining Magazine*, London, September 1929; and 'Uranium and radium in South West England' by Neil Dickinson, in the *Journal of the Plymouth Mineral & Mining Club*, vol.II, September 1980. See also files of Société industrielle du radium: London, PRO, BT31/32141/125420
14 *The Times*, London, 16 March 1915
15 *The Scotsman*, Edinburgh, 18 March, 1915; and *Nature*, London, 25 March 1915
16 *The Times*, London, 24 June and 7 August 1914
17 Files of J.S. MacArthur Ltd, National Archives of Scotland, Edinburgh, BT2/9375
18 The Radium Products Co. and its successor continued in business at several different addresses. As late as 1934, 'radium products' such as Spa Radium Sparklet Syphons – radon gas bulbs for the production of radioactive water – were still advertised. Records of neither company survive
19 'Review of radium hazards and the regulation of the radium industry' by R.M. Pratt, in *Environment International*, vol.19, 1993, pp.475–89
20 'Production of radium in America' by Charles H. Viol, in *Mining & Scientific Press*, San Francisco, 19 September 1914; see also 'Radium, uranium and vanadium in 1914' in *The Mining Journal*, London, 6 March 1915; 'The radium situation' by Warren F. Bleecker, in *Metallurgical and Chemical Engineering*, New York, March 1915; 'The Standard Chemical Co.' by Joseph M. Flannery, in *The Salt Lake Mining Review*, Salt Lake City, 15 April 1915. The best and most comprehensive account of the early history of radium in America is 'Buried treasure to buried waste: the rise and fall of the radium industry' by Edward R. Landa, in *Colorado School of Mines Quarterly*, vol.82, no.2, 1987, Colorado School of Mines Press, Golden, Colorado: a complete volume on the subject
21 Information from the late Mr J. Summers and Mrs J. MacPherson, Balloch

CHAPTER FIVE

1 *The Times*, London, 27 January 1909. Treves was the most famous surgeon of his time, turning a bloody craft into a modern science. He was court physician to three monarchs; a field surgeon in the Boer War; and in retirement became a celebrated travel writer. Joseph Merrick suffered from incurable, grotesque facial and skeletal deformities. He was virtually a hostage in a series of circus freak-shows, and was thought to be an imbecile. Treves rescued him from his wretched situation, arranging permanent accommodation for him in the London Hospital until his death. Under Treves' care, Merrick proved to be highly intelligent and articulate
2 *Nature*, 31 October 1912
3 *The Times*, London, 3 October 1913*7
4 *The Times*, London, 3 October 1913
5 *The Times*, London, 3 October 1913

6 *The Times*, London, 3 October 1913
7 *The Times*, London, 25 April 1914
8 *The Times*, London, 25 April 1914
9 *The Mining Journal*, 29 November 1913, p.1,140
10 *The Mining Journal*, 29 November 1913, p.1,140
11 *The Mining Journal*, 29 November 1913, p.1,140. In personal preference to 'radon emanation', Ramsay unsuccessfully tried to bring into currency the term 'niton' – from the Latin for 'shining' and 'on' from the suffix used in his series of noble gases
12 *The Mining Journal*, 29 November 1913, p.1,140
13 *The Times*, 11 December 1913
14 'Buried treasure to buried waste: the rise and fall of the radium industry' by Edward R. Landa, in *Colorado School of Mines Quarterly*, vol.82, no.2, 1987, p.9
15 *The Mining Journal*, London, 12 July 1913
16 *The Glasgow Herald*, 29 December 1913
17 *Radioactivity in America*, by Lawrence Badash, p.143
18 *The Glasgow Herald*, 31 December 1913
19 *The Times*, 7 January 1914; and *The Glasgow Herald*, 10 January 1914
20 *The Times*, 26 November 1913; British Patent no.19,820 of 1909
21 *The Times*, 7 January 1914; British Patent nos 28,444 and 28,445 of 1913
22 *The Times*, 7 January 1914; and British Patent no. 25,504 of 1910, by Frederick Soddy. (Mesothorium, a decay product of thorium, has two forms, now known as the isotopes radium-228 and actinium-228)
23 *The Glasgow Herald*, 9 January 1914
24 *The Glasgow Herald*, 10 January 1914
25 *The Times*, 10 January 1914
26 Quoted by Charles L. Parsons (Chief of the Division of Mineral Technology, US Bureau of Mines) in 'Our Radium Resources', *Mining Science*, December 1913, pp.319–24.
27 *The Times*, 20 January 1914
28 'Radium production in America' by Charles H. Viol, *Mining & Scientific Press*, 19 September 1914, pp.443–44
29 'Radium production in America' by Charles H. Viol, *Mining & Scientific Press*, 19 September 1914, pp.443–44
30 'The radium situation' by Warren F. Bleecker, in *Metallurgical and Chemical Engineering*, March 1915, pp.143–45
31 'The radium situation' by Warren F. Bleecker, in *Metallurgical and Chemical Engineering*, March 1915, pp.143–45
32 'The first nuclear industry' by Edward R. Landa, in *Scientific American*, November 1982, pp.154–63
33 'Our radium resources' by Charles L. Parsons, in *Mining Science*, December 1913, pp.319–24.
34 *Science*, New York, May 1915
35 *Nature*, 5 November 1903
36 *Scottish Country Life*, March 1917, p.120

37 *The Times*, London, 25 September 1915; and *Nature*, London, 7 October 1915
38 Lecture by T. Thorne Baker, 'The industrial uses of radium' reproduced in the *Journal of the Royal Society of Arts*, London, 16 April 1915
39 Lecture by T. Thorne Baker, 'The industrial uses of radium' reproduced in the *Journal of the Royal Society of Arts*, London, 16 April 1915
40 J.S. MacArthur, 'The radium industry and reconstruction' in *The Mining Journal*, London, 18 January 1919
41 'Buried treasure to buried waste: the rise and fall of the radium industry' by Edward R. Landa, in *Colorado School of Mines Quarterly*, vol.82, no.2, 1987, p.36.
42 'Buried treasure to buried waste: the rise and fall of the radium industry' by Edward R. Landa, in *Colorado School of Mines Quarterly*, vol.82, no.2, 1987, p.36
43 *Atom*, UK Atomic Energy Authority, October 1987

CHAPTER SIX

1 'Buried treasure to buried waste: the rise and fall of the radium industry' by Edward R. Landa, in *Colorado School of Mines Quarterly*, vol.82, no.2, 1987, p.35
2 *The Lancet*, 27 February 1909, p.654
3 Mount Clemens doctor quoted in *Radium Girls*, by Claudia Clark, p.49
4 US Federal Trade Commission records, quoted in *Radium Girls*, by Claudia Clark, p.51
5 The term 'radon' did not come into common use until the late 1920s; until then 'radium emanation' was preferred
6 Report of a meeting in the Pump Room at Bath, in *The Times*, London, 12 March 1912
7 Bath Hot Springs advertising pamphlet
8 'On radium emanations in mineral waters', by T. Pagan Lowe, in *The Lancet*, 20 April 1912, p.10,53
9 'On radium emanations in mineral waters', by T. Pagan Lowe, in *The Lancet*, 20 April 1912, p.10,52
10 Files of the Radium Salt Co. Ltd, Public Record Office, London. BT31/22443/137071
11 'The industrial uses of radium', by T. Thorne Baker, in *Journal of The Royal Society of Arts*, London, 16 April 1915, pp.490–98
12 'The industrial uses of radium', by T. Thorne Baker, in *Journal of The Royal Society of Arts*, London, 16 April 1915, pp.490–98
13 'The industrial uses of radium', by T. Thorne Baker, in *Journal of The Royal Society of Arts*, London, 16 April 1915, pp.490–98
14 *Nature*, London, 9 October 1913
15 *The Times*, London, 8 and 28 November 1913
16 *The Times*, London, 12 December 1913
17 *Journal of the American Medical Association*, 23 December 1911
18 *The Times*, London, 9 February 1914
19 *The Times*, London, 30 March 1914
20 *Journal of the American Medical Association*, April 1927

21 *Radium lost and found*, by R.B. Taft, p.95
22 'Radium hounds' by R.B. Taft, *Scientific American*, vol.160 no.1, 1939, p.8
23 *The Times*, London, 22 and 25 April 1914
24 *Journal of the American Medical Association,* 6 August 1927
25 Gable's promotional material, University of Iowa Library (Special Collections Dept), 1A 52242-1420
26 Account given in *Radioactivity in America: Growth and Decay of a Science* by Lawrence Badash, Johns Hopkins University Press, Baltimore, 1979 (referring to *The New York Times*, January 1914)
27 J.S. MacArthur, 'The radium industry and reconstruction' in *The Mining Journal*, London, 18 January 1919
28 J.S. MacArthur, 'The radium industry and reconstruction' in *The Mining Journal*, London, 18 January 1919
29 *The Times* engineering supplement, London, 20 February 1920
30 *The Times*, London, 12 January 1921
31 *Journal of the American Medical Association*, 4 February 1933
32 Account quoted in *A History of X-rays and Radium* by Richard F. Mould, IPC Business Press, London, 1980; see also *Multiple Exposures* by Catherine Caufield, Secker & Warburg, London, 1989
33 'Buried treasure to buried waste: the rise and fall of the radium industry' by Edward R. Landa, in *Colorado School of Mines Quarterly*, vol.82, no.2, 1987, p.40

CHAPTER SEVEN

1 *The Manufacturer and Builder*, New York, December 1882, p.286
2 *The Manufacturer and Builder*, New York, March 1878, p.67
3 Balmain's Luminous Paint was patented in the UK in 1877 and in the USA in 1882
4 *Scientific American*, Supplement no.497, 11 July 1885
5 'Buried treasure to buried waste: the rise and fall of the radium industry' by Edward R. Landa, in *Colorado School of Mines Quarterly*, vol.82, no.2, 1987
6 *Radioactivity in America – Growth and Decay of a Science*, by Lawrence Badash, pp.34–48
7 *Radium in Humans, a Review of US Studies*, R.E. Rowland, p.11
8 *Radium and other radioactive substances*, by William J. Hammer, 1904, p.18
9 *Radium and other radioactive substances*, by William J. Hammer, 1904, p.18
10 Smithsonian Institution: http://americanhistory.si.edu/archives/d8069c.htm
11 'The story of radium', in *Scientific American*, 24 April 1920, p.454
12 'Self-luminous paint' in *Engineering & Mining Journal*, New York, November 1916; 'Radioactive luminous materials' by Wallace Savage, in *Chemical and Metallurgical Engineering*, New York, 28 September 1918; 'Radioactivity and some practical applications' by C.W. Davis, in *Mining & Scientific Press*, San Francisco, 12 February 1921; and 'The application of radium in warfare', a paper by Charles Viol and Glenn Kammer (of Standard Chemical Co.) at the 32nd AGM of the American Electro-chemical Society, Pittsburgh, 5 October, 1917

13 Advisory Committee for Aeronautics, January 1916; PRO: DSIR 23/630
14 *Mineral Industry* ('Radium' by Robert M. Keeney), New York, 1916, p.641
15 *Salt Lake Mining Review*, 15 November 1916
16 Ingersoll Watches advertisement, 1917
17 *Popular Science Monthly*, New York, December 1918
18 *Popular Science Monthly*, New York, February 1918
19 *Radioactivity in the Environment*, by Ronald L. Kathren, p.17
20 See *Radioactivity in America – Growth and Decay of a Science* by Lawrence Badash, Johns Hopkins University Press, Baltimore, 1979
21 *Madame Curie*, by Eve Curie, Da Capo edition, p. 232–23
22 'Buried treasure to buried waste: the rise and fall of the radium industry' by Edward R. Landa, in *Colorado School of Mines Quarterly*, vol.82, no.2, 1987, p.45
23 *Engineering & Mining Journal*, 30 December 1916
24 'Radioactivity and some practical applications', by C.W. Davis, in *Mining & Scientific Press*, 12 February 1921, p.231
25 Ministry of Munitions Order, printed in *The Times*, London, 14 August 1918. See also PRO: CAB 40/98
26 Dissolved companies' files, PRO, BT31 series
27 'The radium industry and reconstruction', by J.S. MacArthur in *The Mining Journal*, London, 18 January 1919
28 John S. MacArthur, notebook: the MacArthur Collection, Balliol College, Oxford University. Library catalogue ref. Box1/VI/1
29 John S. MacArthur, notebook: the MacArthur Collection: Balliol College, Oxford University. Library catalogue ref. Box1/VI/1.
30 'The radium industry and reconstruction', by J.S. MacArthur in *The Mining Journal*, London, 18 January 1919
31 *The Times*, London, 18 January 1919
32 See www.mindat.org/loc-1516.html
33 'The recovery of radium from luminous paint', by A.G. Francis (Government Laboratory, Clement's Inn Passage, London), in *Journal of the Society of Chemical Industry*, London, 31 March 1922
34 J.S. MacArthur, obituary notice, *Nature*, London, 25 March 1920
35 'The South Terras radium deposit, Cornwall' by T. Robertson and H. Dines, in *The Mining Magazine*, London, September 1929. See also www.dangerouslaboratories.org/rcw3.html
36 *The Deadly Element: the Story of Uranium*, by Lennard Bickel, pp.48–51
37 Information on Belgian operations from *Radium, production, general properties, therapeutic applications, apparatus, etc.*, Union minière du Haut-Katanga, Brussels, no date (*c.*1928); and 'The origin and early development of the Belgian radium industry' by A. Adams, in *Environment International*, vol.19, 1993, pp.491–501.
38 *Nature and magnitude of the problem of spent radiation sources,* IAEA-TECDOC-620 (1991)
39 'Buried treasure to buried waste: the rise and fall of the radium industry' by Edward R. Landa, in *Colorado School of Mines Quarterly*, vol.82, no.2, 1987, p.26. See also

the narrative accompanying records of the US Radium Corporation in the US Library of Congress, Historic American Engineering Record, (HAER NJ 007-ORA,3), available via the Library of Congress website http://memory.loc.gov; and see also 'Early uranium mining in the United States', a paper by F.J. Hahne for the Uranium Institute in London, September 1989, at www.world-nuclear.org/usumin.htm
40 Letter in *The Times*, London, 13 December 1919

CHAPTER EIGHT

1 *New York World*, 3 August 1903
2 *New York World*, 3 August 1903
3 *Archives of the Roentgen Ray*, no.18, April 1914
4 *Radioactivity in America*, by Lawrence Badash, p. 27.
5 *Radioactivity in the Environment,* by Ronald L. Kathren, p.16
6 *Nuclear Fear,* by Spencer R. Weart, p.50
7 'Buried treasure to buried waste: the rise and fall of the radium industry' by Edward R. Landa, in *Colorado School of Mines Quarterly*, vol.82, no.2, 1987, p.36
8 'Buried treasure to buried waste: the rise and fall of the radium industry' by Edward R. Landa, in *Colorado School of Mines Quarterly*, vol.82, no.2, 1987, p.35
9 Revigator brochure, quoted at the website of Oak Ridge Associated Universities, www.orau.gov/ptp/collection/quackcures/revigat.htm
10 *Journal of the American Medical Association*, 21 November 1925, p.1,659
11 *Journal of the American Medical Association*, 21 November 1925, p.1,659
12 *Journal of the American Medical Association*, 21 November 1925, p.1,658
13 Oak Ridge Associated Universities, www.orau.gov/ptp/collection/quackcures/quackcures.htm
14 British Patent no.27,925 of 1909
15 British patent no.5,182 of 1913
16 *Radio Times*, June 1932
17 *Radioactivity in the Environment,* by Ronald L. Kathren, p.14
18 Report by Harrop, Wolstenholme and Allen, in *Chemistry in Britain*, May 1992
19 Oak Ridge Associated Universities, www.orau.gov/ptp/collection/quackcures/quackcures.htm
20 'Buried treasure to buried waste: the rise and fall of the radium industry' by Edward R. Landa, in *Colorado School of Mines Quarterly*, vol.82, no.2, 1987, p.37
21 Oak Ridge Associated Universities, www.orau.gov/ptp/collection/quackcures/radend.htm
22 *Journal of the American Medical Association*, vol.88, 1927, p.343
23 'Radium as a patent medicine' in *Journal of the American Medical Association*, 16 April 1932, p.1, 398
24 'Radithor and the era of mild radium therapy' by Roger M. Macklis in *Journal of the American Medical Association*, 1 August 1990, p.618
25 Unidentified newspaper of 28 March 1925, quoted in *Radium Girls*, by Claudia Clark, p.172

26 Attorney Robert H. Winn, quoted in 'Radithor and the era of mild radium therapy' by Roger M. Macklis in *Journal of the American Medical Association*, 1 August 1990, p.616

27 For information on Bailey, see 'Radium as a patent medicine' in *Journal of the American Medical Association*, 16 April 1932; *Popular Science Monthly*, New York, July 1932; 'The great radium scandal' by Roger M. Macklis in *Scientific American* vol.269, no.2, August 1993; and 'Radithor and the era of mild radium therapy' by Roger M. Macklis in *Journal of the American Medical Association*, 1 August 1990. See also *Multiple Exposures* by Catherine Caufield, 1989; 'A history of pitchblende' by Dr J.R. Morgan, in *Atom*, UKAEA, Harwell, March 1984; and *Radioactivity in the Environment* by Ronald L. Kathren, 1984

28 C. Everett Field, quoted in *Radium Girls*, by Claudia Clark, p.62

29 'Radium as a patent medicine' in *Journal of the American Medical Association*, 16 April 1932, p.1,399

30 Bailey, quoted in 'The great radium scandal' by Roger M. Macklis, in *Scientific American*, vol.269 no.2, August 1993, p.83

31 'Radithor and the era of mild radium therapy' by Roger M. Macklis in *Journal of the American Medical Association*, 1 August 1990, p.618

CHAPTER NINE

1 See the narrative accompanying the records of the US Radium Corporation in the US Library of Congress, Historic American Engineering Record (HAER NJ 007-ORA,3-), available via the Library of Congress website http://memory.loc.gov

2 S.A. von Sochocky, 'Can't you find the keyhole?', in *The American Magazine*, New York, January 1921

3 *Radium Girls: Women and industrial health reform, 1910–1935*, by Claudia Clark, pp.14–17

4 'Public Health Assessment, Montclair/West Orange Radium Site', New Jersey Department of Health: see website http://atsdr1.atsDrcdc.gov:8080/HAC/PHA/montclair/mon_p1.html

5 Dr Robley Evans, 'Radium poisoning: a review of our present knowledge', in *The American Journal of Public Health*, vol.23, 1933. Evans began this study in 1932 as a student, and continued into the 1970s as head of the Radium Dial Painter Study Group at the Massachusetts Institute of Technology

6 'Some unrecognised dangers in the use and handling of radioactive substances', by Harrison S. Martland, P. Conlon, and J. Knef, in *Journal of the American Medical Association*, 5 December 1925, p.1,770

7 'Occupational poisoning in manufacture of luminous watch dials', by Harrison S. Martland, in *Journal of the American Medical Association*, 9 February 1929, p.468

8 Editorial in *New York World*, 19 May 1928 (quoted in *Radium Girls: Women and industrial health reform, 1910–1935*, by Claudia Clark, p.133)

9 'Some unrecognised dangers in the use and handling of radioactive substances', by Harrison S. Martland, P. Conlon and J. Knef, in *Journal of the American Medical Association*, 5 December 1925

10 Hoffman, 'Radium (mesothorium) necrosis', in *Journal of the American Medical Association*, 26 September 1925, p.965
11 Statements given by family members in the 1986 documentary film *Radium City* by Carole Langer
12 Sources for the story of the dial-painters include: S.A. von Sochocky, 'Can't you find the keyhole?', in *The American Magazine*, New York, January 1921; Dr Robley Evans, 'Radium poisoning: a review of our present knowledge', in *The American Journal of Public Health*, vol.23, 1933; 'Some unrecognised dangers in the use and handling of radioactive substances', by Harrison S. Martland, P. Conlon and J. Knef, in *Journal of the American Medical Association*, 5 December 1925; 'Occupational poisoning in manufacture of luminous watch dials', by Harrison S. Martland, in *Journal of the American Medical Association*, 9 February 1929; *Radium Girls: Women and industrial health reform, 1910–1935* by Claudia Clark; 'Doomed to die – and they live!' in *Popular Science Monthly*, New York, July 1929; 'Radium deals death in hands of quacks', in *Popular Science Monthly*, New York, July 1932; 'Health aspects of radium dial painting', by Schwarz *et al.*, in *Journal of Industrial Hygiene*, Baltimore, September 1933; 'Mortality among women employed before 1930 in the U.S. radium dial-painting industry', by Polednak, Stehney and Rowland, in *American Journal of Epidemiology*, Baltimore, March 1978; *Radium in Humans, a Review of US Studies*, by R.E. Rowland, Argonne National Laboratory, 1994; 'Buried treasure to buried waste: the rise and fall of the radium industry' by Edward R. Landa in *Colorado School of Mines Quarterly*, vol.82, summer 1987; the US Library of Congress, Historic American Engineering Record (HAER NJ 007-ORA,3), available via the Library of Congress website http://memory.loc.gov; *Multiple Exposures: Chronicles of the Radiation Age* by Catherine Caufield; and the American documentary film *Radium City* by Carole Langer
13 Public statement by the Radium Dial Co., published in the *Daily Republican Times*, 7 June 1928
14 Letter from US Radium president, 18 June 1928, quoted in *Radium in Humans, a Review of US Studies*, by R.E. Rowland, p.13
15 *Radium Poisoning*, US Department of Labor report of 1912, p.1,231, quoted in 'Buried treasure to buried waste: the rise and fall of the radium industry', by Edward R. Landa, *Colorado School of Mines Quarterly*, vol.82, summer 1987, p.24
16 'Radioactive material: an industrial hazard?' By Dr Frederick B. Flinn, in *Journal of the American Medical Association*, 18 December 1926, p.2,079
17 'Occupational poisoning in manufacture of luminous watch dials', by Harrison S. Martland, in *Journal of the American Medical Association*, 9 February 1929
18 'Occupational poisoning in manufacture of luminous watch dials', by Harrison S. Martland, in *Journal of the American Medical Association*, 9 February 1929
19 'Occupational poisoning in manufacture of luminous watch dials', by Harrison S. Martland, in *Journal of the American Medical Association*, 9 February 1929
20 'Occupational poisoning in manufacture of luminous watch dials', by Harrison S. Martland, in *Journal of the American Medical Association*, 9 February 1929
21 'Occupational poisoning in manufacture of luminous watch dials', by Harrison S. Martland, in *Journal of the American Medical Association*, 9 February 1929

22 *The Times*, 27 November 1962, p.10
23 'Occupational poisoning in manufacture of luminous watch dials', by Harrison S. Martland, in *Journal of the American Medical Association*, 9 February 1929
24 *Scientific American*, March 1947, p.134
25 'Review of radium hazards and regulation of the radium industry' by R.M. Pratt, in *Environment International*, vol.19, 1993, pp.475–89
26 US Library of Congress, Historic American Engineering Record (HAER NJ 007-ORA,3), available via the Library of Congress website http://memory.loc.gov/
27 US Library of Congress, Historic American Engineering Record (HAER NJ 007-ORA,3), available via the Library of Congress website http://memory.loc.gov/
28 See www.phillyspy.com
29 'Some factors contributing to the internal radiation hazard in the radium luminising industry', by Duggan and Godfrey (Radiological Protection Service) in *Health Physics*, vol.13, Pergamon Press, Oxford, 1967
30 *Radium in Humans, a Review of US Studies*, by R.E. Rowland, Argonne National Laboratory, 1994, p.2
31 *Adventures in the Atomic Age,* by Glenn Seaborg and Eric Seaborg
32 Libby, quoted in *Project Sunshine and the Slippery Slope*, by Sue Rabbit Roff, Dundee University Medical School. For a selection of original documents, see also www.gwu.edu/~nsarchiv/radiation/dir/mstreet/commeet/meet15/brief15/tab_d/tab_d.html
33 See 'The verdict: no harm, no foul', by Danielle Gordon, in *The Bulletin of the Atomic Scientists* vol.52 no.1, January/February 1996
34 Hamilton memo quoted by Eileen Welsome in *The Plutonium Files*, p.321
35 Dr Alice Stewart in *Deadly Experiments*, Channel 4 Television, 6 July 1995
36 Dr Alice Stewart in *Deadly Experiments*, Channel 4 Television, 6 July 1995
37 *Bulletin of the Atomic Scientists*, vol.52, no.1, January/February 1996; and news reports in *The Guardian*, London, 30 December 1993, 1 and 12 January 1994, and 6 July 1995; Channel 4 television documentary, *Deadly Experiments*, broadcast in July 1995; see also internet website of the US government's Office of Human Radiation Experiments, www.ohre.doe.gov; and the report 'British nuclear guinea-pigs' by CND at mannet.mcb.net/cnd/radexpts/report.htm
38 *The Herald, Glasgow* 12 March 2004

CHAPTER TEN

1 *Journal of Industrial & Engineering Chemistry*, August 1926
2 *Journal of Industrial & Engineering Chemistry*, August 1926
3 'Buried treasure to buried waste', by Edward R. Landa, in *Colorado School of Mines Quarterly*, vol.82 no.2, summer 1987, p.27
4 *Journal of the American Medical Association*, 28 February 1925, p.687
5 *The Lancet*, 20 April 1929, pp.850–51
6 *The Lancet*, 5 April 1930, p.783

7 *The Lancet*, 20 April 1929, p.850
8 *The Lancet*, 4 May 1929, pp.929–30
9 *The Lancet*, 4 May 1929, p.929
10 'Radium treatment of cancer in France and Belgium' by Donald J. Armour and Henry Souttar, in *The Lancet*, 9 February 1929, pp.299–301
11 *Journal of the American Medical Association*, 17 October 1925, p.1,234
12 *Journal of the American Medical Association*, 30 April 1927, p.1,429
13 Account of LaBine's discovery in *Radium*, by Lilian Holmes Strack, 1941
14 Northwest Territories History of Exploration and Development, www.gov.nt.ca/RWED/mog/minerals.mins_history.htm
15 'Radium from the Canadian Arctic', by Marcel Pochon, in *Engineering & Mining Journal*, September 1937, pp.39–41
16 *Radium* by Lilian Holmes Strack, pp.25–34; 'Radium from the Arctic' by H.C. Parmelee in *Engineering & Mining Journal*, April 1938, pp.31–35; and 'Last frontier' by Leslie Roberts in 'Canadian Mining Journal', July 1937, pp.361–68
17 Statement by American Chemical Society, 'Radium strike in Canadian wilds', by Charles McLeod in *Popular Science Monthly*, April 1932, pp.17–20
18 'Memoirs of a radiochemist' reproduced courtesy of Prof. Donald R. Wiles, Ottawa; at www.carleton.ca/~dwiles/radiochemistry/memoires.htm
19 'Review of radium hazards and regulation of the radium industry' by R.M. Pratt, in *Environment International*, vol.19, 1993, pp.475–89
20 'Hunting uranium around the world' by Robert D. Nininger, in *National Geographic Magazine*, October 1954, pp.533–56
21 Cameco Corporation: www.cameco.com/uranium_101
22 World uranium deposits, OECD/IAEA
23 'The origin and early development of the Belgian radium industry' by A. Adams, in *Environment International*, vol.19, 1993, pp.491–501
24 'Review of radium hazards and regulation of the radium industry' by R.M. Pratt, in *Environment International*, vol.19, 1993, pp.475–89
25 *The Making of the Atomic Bomb*, by Richard Rhodes, p.427
26 *The Making of the Atomic Age*, by Alwyn McKay, p.36
27 *Dark Sun*, by Richard Rhodes, p.130
28 'Hunting uranium around the world' by Robert D. Nininger, in *National Geographic Magazine*, October 1954, pp.533–56
29 Gordon Dean, AEC Chairman, 'Report on the atom', quoted in *Multiple Exposures* by Catherine Caufield, p.75
30 *People* magazine, 15 December 1954, quoted in *Uranium Fever*, by Raymond and Samuel Taylor, p.3
31 *Uranium – Where it is and how to find it*, by Proctor, Hyatt and Bullock, p.1
32 *Uranium Fever*, by Raymond and Samuel Taylor, p.116
33 *Dark Sun*, by Richard Rhodes, p.214
34 Secret 'Report on former radium plant at Dumbarton [sic] of 11 December 1948, by the Chief Geologist, Atomic Energy Division of the Geological Survey and Museum to the Atomic Energy Secretariat of the Ministry of Supply'

35 Glenn Seaborg, quoted from *Harper's Magazine*, 1970. The suggestion had been made in *Scientific American* (1947, p.63) that uranium had unique energy powers, and that energy 'is a far more logical basis of economic value than any possessed by the precious metals'
36 Oliphant, quoted in *The Deadly Element, the story of uranium*, by Lennard Bickel, p.287

CHAPTER ELEVEN

1 'Nothing clean about "cleanup" ', by Linda Rothstein in *Bulletin of the Atomic Scientists*, May/June 1995
2 *The Guardian*, 18 August 1993
3 *The Prague Post*, 8 January 2003
4 See website for Jáchymov spa facilities: www.laznejachymov.cz
5 'Radiological aspects of former mining activities in the Saxon Erzgebirge, Germany', by Gert Keller, in *Environment International*, 1993, pp.449–54
6 See the excellent website on the reclamation of uranium mining sites across the world at www.antenna.nl/wise/uranium/ ; and www.antenna.nl/wise/uranium/uwis.html for 'Uranium mining in Eastern Germany – The Wismut Legacy' by Peter Diehl. See also 'The legacy of uranium mining in Central and Eastern Europe – a view from the European Union' by Simon Webster, Directorate-General Environment, European Commission, Brussels
7 '1990 Recommendations of the ICRP', p.71, see ICRP website 'Summary of Recommendations' at www.icrp.org/educational_area.asp
8 'The scientific development of a former gold mine near Badgastein, Austria, to the therapeutic facility "Thermal Gallery" ' by J. Pohl-Ruling, in *Environment International*, 1993, pp.455–65; 'A history of pitchblende', by Dr J.R. Morgan, in *Atom*, (UKAEA) Harwell, March 1984; and 'Buried treasure to buried waste: the rise and fall of the radium industry', by Edward R. Landa, in *Colorado School of Mines Quarterly*, vol.82, summer 1987
9 'Badgastein – A natural spectacle in 8 scenes', Brochure, Badgastein Spa and Tourism Association, 1991 (see also www.gastein.com)
10 Gastein Healing Gallery, www324.s5.brauser.at/healing.htm
11 Free Enterprise Health Mine website, www.radonmine.com
12 *Life*, Chicago, 28 July 1952
13 *Le Monde*, 24 March 2004; Associated Press, 31 May and 1 June 2004
14 'Cleaning up the radioactive contamination on the Umicore sites at Olen and in the vicinity': see www.um.be/comrel/en/group/radioactiviteit/radioactivVisieE.htm
15 'Echoes of the atomic age' by Andrew Nikiforuk, in *Calgary Herald*, 14 March 1998
16 See Canadian Low-Level Radioactive Waste Management website, www.llrwmo.org/en/porthope/history.html
17 'Cleanup of radioactive contamination at Port Hope', by Pearl Marshall, in *Nuclear Engineering International*, July 1976
18 See the website of the US Agency for Toxic Substances and Disease Registry, http://atsdr1.atsDrcdc.gov:8080/HAC/PHA/montclair/mon_p1.html; also www.epa.gov/region02/superfund/npl/0200997c.htm

19 See www.phillyspy.com
20 *Radium City*, US documentary by Carole Langer, 1986
21 See the website of the US Agency for Toxic Substances and Disease Registry, http://atsDrcdc.gov/HAC/PHA/ottawa/ott_p1.html
22 'Will these lands ne'er be clean?' by Don Charles, in *New Scientist*, 24 June 1989; see also 'America's clean-up', in *International Mining*, 24 October 1990; and *Multiple Exposures: Chronicles of the Radiation Age* by Catherine Caufield, 1989
23 UMTRA website, www.doeal.gov/oepm/factgrj.htm
24 www.defra.gov.uk/rwmac/reports/mod/p5_1.htm
25 *Glasgow Herald*, 9 June 1997
26 'Historic practices in the UK which have utilised radioactive materials', DEFRA, DETR/RAS/00.005; see website www.defra.gov.uk/environment/radioactivity/research/complete/historic/index.htm
27 *The Guardian*, 9 December 1999
28 Tritium is a radioactive isotope of hydrogen. In nuclear weapons, because of its short half-life of twelve years, it has to be regularly removed and replaced. It is no longer produced in the USA, and the UK is the major supplier. There has been increasing concern over its potential for causing biological damage
29 *The Observer*, 30 April 2000
30 *The Guardian*, 3 March 1988
31 *The Sunday Herald*, 5 December 1999 and 6 August 2000
32 *Tritium in Scottish Landfill Sites* by Hicks, Wilmot & Bennett, May 2000; at www.sepa.org.uk/pdf/publications
33 *The Observer*, 18 July 1999
34 www.amersham.com
35 *The Guardian*, 22 February 1989
36 House of Commons written reply, March 1988
37 'Assessment of the implications of radium contamination of Dalgety Bay Beach and foreshore', University of Aberdeen and Auris Environmental Ltd for the Scottish Office Central Research Unit, 1996
38 Report following site visit in March 2000 by the Radioactive Waste Management Advisory Committee: www.defra.gov.uk/rwmac/reports/mod
39 Parish of Runcorn Rate-books, 6 October 1910 (which related to premises liable for rates prior to March 1911)
40 From a report by Halton Borough Council sent to the author in March 1996
41 Report of 24 August 1961 by the Chemical Inspector, Department of Health for Scotland
42 Letter of 24 October 1961 from the Radiation Protection Service to the Department of Health for Scotland
43 ^{210}Pb, 6,500 Bq/kg^{-1}; ^{226}Ra, 2,200 Bq/kg^{-1}; ^{214}Bi, 1,100 Bq/kg^{-1}; ^{214}Pb, 1,100 Bq/kg^{-1}
44 'Report on Radium Works site, Dalvait, Balloch' by Dr J. Gemmill and Dr P. Smith, Strathclyde Regional Council Department of the Regional Chemist, September 1991

45 'Report on Radium Works site, Dalvait, Balloch' by Dr J. Gemmill and Dr P. Smith, Strathclyde Regional Council Department of the Regional Chemist, September 1991
46 See *Radiation Research*, vol.146, p.247; and *New Scientist*, vol.156, no.2103, 11 October 1997
47 House of Commons Environment Committee, 1st Report on Contaminated Land, vol.1. (House of Commons Paper 170-1, HMSO, 1990)
48 *The Guardian*, 31 August 2004
49 *The Guardian*, 14 April 2004; see also website of the Committee on Radioactive Waste Management, www.corwm.org.uk
50 *The Observer*, 7 July 1991

EPILOGUE

1 *Popular Science Monthly*, no.57, 1900, pp.318–22, quoted in Lawrence Badash, 'Radium, radioactivity and the popularity of scientific discovery, *Proceedings of the American Philosophical Society*, no.122, 1978, pp.145–54
2 Frederick Soddy, *The Interpretation of Radium*, 3rd edition, 1912, p.251
3 Waldemar Kaempffert, *New York Times Magazine*, January 1934, quoted in *Nuclear Fear*, by Spencer R. Weart, p.13
4 *New Scientist*, 20 March 2004
5 Food Standards Agency, FSIS 67/04, September 2004
6 C. Northcote Parkinson, *Parkinson's Law*, Chapter 10

BIBLIOGRAPHY

BOOKS

Badash, Lawrence. *Radioactivity in America: Growth and Decay of a Science.* Johns Hopkins University Press, Baltimore, 1979

Bauer, Georg (Agricola). *De re metallica*, translation of 1556 edition by Herbert Clark Hoover and Lou Henry Hoover, London 1912

Bertell, Rosalie. *No Immediate Danger: Prognosis for a radioactive Earth.* The Women's Press, London, 1985

Bickel, Lennard. *The Deadly Element.* Stein & Day, New York, 1979

Bruce, Robert V. *Bell: Alexander Graham Bell and the Conquest of Solitude.* Cornell University Press, Ithaca, 1973

Caufield, Catherine. *Multiple Exposures.* Secker & Warburg, London, 1989

Clark, Claudia. *Radium Girls: Women and Industrial Health Reform, 1910–1935.* University of North Carolina Press, 1997

Curie, Eve. *Madame Curie* (translated by Vincent Sheean). Da Capo Paperback, New York, 1986 (first published by Doubleday & Co., 1937)

Curie, Marie. *Radioactive Substances.* Dover Publications, New York. (This is Curie's doctoral dissertation, published as *Recherches sur les substances radioactives* in 1903; the English translation, of which this is an unabridged edition, was published in the journal *Chemical News* and by Van Nostrand, New York, in 1904)

Curie, Marie. *Pierre Curie* (and autobiographical notes). Macmillan, New York, 1923

Doerner, H.A. 'Metallurgy of radium and uranium', in *Handbook of Non Ferrous Metallurgy.* McGraw-Hill, New York, 1945

Gould, Jay M. and Goldman, Benjamin A. *Deadly Deceit: Low-level Radiation, High-level Cover-up.* Four Walls Eight Windows, New York, 1991

Hall, Jeremy. *Real Lives, Half Lives; Tales from the Atomic Wasteland.* Penguin, 1996

Hammer, William J. *Radium and other Radioactive Substances.* Keegan Paul, Trench,

Trubner & Co., London, 1904
IAEA. *The Environmental Behaviour of Radium*. Vienna, 1990
Kathren, Ronald L. *Radioactivity in the Environment: Sources, Distribution and Surveillance*. University of Washington, 1984
Keller, Alex. *The Infancy of Atomic Physics – Hercules in his Cradle*. Clarendon Press, 1983
Lambert, Dr Barrie. *How Safe is Safe? Radiation Controversies Explained*. Unwin Paperbacks, London, 1990
Mackay, Alan L. (ed.). *A Dictionary of Scientific Quotations*. Adam Hilger, Bristol, 1991
Martin, Alan, and Harbison, Samuel A. *An Introduction to Radiation Protection* (3rd ed.). Chapman and Hall, London, 1990
Mould, Richard F. *A History of X-Rays and Radium*. IPC Business Press, London, 1980
Niewenglowski, G.H. *Les rayons X et le radium*. Librairie Hachette, Paris, 1924
Pasachoff, Naomi. *Marie Curie and the Science of Radioactivity*. Oxford University Press, 1996
Pflaum, Rosalind. *Grand Obsession: Madame Curie and her World*. Doubleday, New York, 1989
Proctor, P.D., Hyatt, E.P., and Bullock, K.C. *Uranium: Where it is and how to find it*. Eagle Rock Publishers, Salt Lake City, 1954
Quinn, Susan. *Marie Curie, a Life*. Heinemann, London, 1995
Rhodes, Richard. *The Making of the Atomic Bomb*. Penguin Books edition, London, 1988
– . *Dark Sun: the Making of the Hydrogen Bomb*. Touchstone edition, New York, 1996
Rowland, R.E. *Radium in Humans, a review of US Studies*. Argonne National Laboratory (ANL/ER-3 UC-408), 1994
Sharp, Robert R. *Early Days in Katanga*. Private imprint, Bulawayo, 1956
Shaw, George Bernard. *The Doctor's Dilemma*
Smith, E.E. *Radiation Science at the National Physical Laboratory, 1912–1955*. HMSO
Strack, Lilian Holmes. *Radium, A Magic Mineral*. Harper & Brothers, New York, 1941
Sumner, David, Wheldon, Tom, and Watson, Walter. *Radiation Risks*. Tarragon Press, Glasgow, 1991
Taft, R.B. *Radium Lost and Found*. Walker, Evans & Cogswell, Charleston, 1946
Taylor, Raymond W. and Samuel W. *Uranium Fever*. Macmillan, New York, 1970
Travers, Morris W. *The Life of Sir William Ramsay KCB, FRS*. Edward Arnold Ltd, London, 1956
Trombley, Stephen. *Sir Frederick Treves, the Extra-ordinary Edwardian*. Routledge, London, 1989
Union minière du Haut-Katanga. *Radium, production, general properties, therapeutic applications, apparatus, etc*. Brussels, no date (*c*.1928)
Weart, Spencer R. *Nuclear Fear: A History of Images*. Harvard University Press, 1988
Welsome, Eileen. *The Plutonium Files, America's secret medical experiments in the Cold War*. The Dial Press, Random House, New York, 1999
Zieroth, Dale. *Nipika, a Story of Radium Hot Springs*. Minister of Supply and Services,

Hull, Quebec, Canada, 1978

PERIODICALS

Archives of the Roentgen Ray
American Journal of Clinical Medicine
American Journal of Epidemiology, Baltimore
American Journal of Public Health
American Magazine, New York
American Medicine
Atom, United Kingdom Atomic Energy Authority
Bulletin of the Atomic Scientists
Canadian Mining Journal
Century Magazine
Chemical & Metallurgical Engineering, New York
Chemistry in Britain, London
Colorado School of Mines Quarterly, summer 1987: 'Buried treasure to buried waste: the rise and fall of the radium industry', by Edward R. Landa
The Cornish Guardian
Engineering, London
Engineering & Mining Journal, New York
Environment International, special edition on the production and use of radium, vol.19, 1993
Le Figaro, Paris
Le Radium, Paris
The Glasgow Herald, Glasgow
Health Physics, Oxford
Journal of the American Medical Association
Journal of Industrial & Engineering Chemistry, Easton, Pennsylvania
Journal of Industrial Hygiene, Baltimore
Journal of Plymouth Mineral and Mining Club
Journal of the Royal Society of Arts
Journal of the Society of Chemical Industry, London
The Lancet, London
Manufacturer and Builder, New York
Metallurgical & Chemical Engineering, New York
Mining and Engineering World, Chicago
The Mining Journal, London
The Mining Magazine, London
Mining Science
Mining & Scientific Press, San Francisco
Nature, London
New Scientist, London
New York World, New York

Popular Science Monthly, New York
Proceedings of the Royal Society, London
Project Sunshine and the Slippery Slope, Sue Rabbit Roff, Dundee University Medical School
Punch, London
Radio Times, London
Salt Lake Mining Review, Salt Lake City
Science, New York
Scientific American
Scottish Country Life
The Times, London
Transactions, US Bureau of Mines

WEBSITES

Database of mines by Heather Coleman (1999): www.dawnmist.demon.co.uk/minedata.htm
Uranium Institute, London: www.uilondon.org/ushist.html and www.world-nuclear.org/usumin.htm
Smithsonian Institution, Washington: http://americanhistory.si.edu/archives
Cornish mine sites: www.dangerouslaboratories.org
Oak Ridge Associated Universities: www.orau.gov/ptp/collection/quackcures
Public Health Assessment, Montclair, West Orange Radium Site: http://atsdr1.atsDrcdc.gov:8080/HAC/PHA/montclair/mon_p1.html
Misc. documents: www.gwu.edu/~nsarchiv/radiation/dir/mstreet/commeet/meet15/brief15/tab_d/tab_d.html
Columbia Journalism Review: http://archives.cjr.org/year/94/2/radiation.asp
US government Office of Human Radiation Experiments: www.ohre.doe.gov
British Nuclear Guinea-pigs: http://mannet.mcb.net/cnd/radexpts/report.htm
Northwest Territories, History of Exploration & Development: www.gov.nt.ca/RWED/mog/minerals.mins_history.htm
US Library of Congress: records of the US Radium Corporation, in the Historic American Engineering Record: http://memory.loc.gov

FILMS AND TELEVISION

Radium City, US documentary by Carole Langer, 1986
Deadly Experiments, UK television documentary by Twenty Twenty TV for Channel Four, broadcast on 6 July 1995 (produced & directed by John Brownlow and Joseph Bullman)

INDEX